# SCANNING NATURE

A look at some of the smaller
components of our natural environment as
revealed by the scanning electron microscope

*Compiled by*

# D. CLAUGHER

BRITISH MUSEUM (NATURAL HISTORY)

CAMBRIDGE UNIVERSITY PRESS

*Note*
Unless otherwise stated, photographs that are part of a series with the same heading, e.g. 'Ant', are all of the same species, which is fully named in the caption of the first photograph in the series.

Published by the British Museum (Natural History), London, and
the Press Syndicate of the University of Cambridge,
The Edinburgh Building, Shaftesbury Road, Cambridge CB2 2RU
32 East 57th Street, New York, NY 10022, USA
296 Beaconsfield Parade, Middle Park, Melbourne 3206, Australia

Library of Congress Catalogue card number: 83-5155

British Library Cataloguing in Publication Data

Scanning nature
   I. Cells      2. Scanning electron microscope
   I. Claugher, D.
   574.8'7      QH585

ISBN 0 521 25705 0  Hardback
ISBN 0 521 27664 0  Limp

Printed in Great Britain by
BAS Printers Ltd, Stockbridge, Hampshire

# Scanning nature

# Contents

# Introduction

The smallest particle that can be resolved by the unaided human eye, given normal light and a contrasting background, is about a twentieth of a millimetre in diameter, but in practice anything less than a tenth of a millimetre in diameter will not be readily detected in ordinary circumstances. The world is full of natural objects of such microscopic size and smaller. Some, such as pollen grains and insect eggs, are ephemeral stages of larger and easily visible plants and animals, but there are many thousands of organisms that never exceed these dimensions at any stage of their life cycle. Some of these, such as certain mites, are just as complex in their bodily structure as most larger animals, both externally and internally, with legs, eyes and other sensory organs, and with alimentary, respiratory, excretory, reproductive and nervous systems. Others, the so-called micro-organisms, are superficially much simpler, comparable to a single cell in a multicellular animal or plant. However, since most of the bodily functions are performed within the bounds of a single cell it is not surprising to find that even micro-organisms are enormously complex and diverse in their structure.

The scanning electron microscope, which first became generally available in 1965, is invaluable in revealing and displaying the surface structure of natural objects, large and small. A sharply focused beam of electrons, in a vacuum, is scanned across the surface of the specimen. This causes secondary electrons to be emitted by the specimen in numbers depending upon its surface topography. These secondary electrons are detected and used to build a picture of the surface contours of the specimen on a cathode-ray display that can then be photographed. A resolution of less than 10 nanometres can be achieved (i.e. two points less than a hundred-thousandth of a millimetre apart can be distinguished), allowing an effective range of magnifications up to about 100 000. However, because of the good depth of focus that can be achieved, it is especially useful for displaying and photographing specimens even at low magnifications that are well within the power of conventional light microscopes.

The photographs presented here have mostly been taken at low magnifications, generally less than x 1000, and have been chosen to demonstrate the complexity and beauty that can be revealed in familiar animals and plants as well as in some less familiar subjects. Many of these subjects can readily be collected and observed by anyone with access to a conventional light microscope although photographs of the quality presented here can only be achieved with the scanning electron microscope.

The scanning electron microscope is playing an increasingly important role in research into natural diversity. By revealing significant differences between superficially similar species of plants and animals — differences that may be of profound importance for the organisms themselves — it allows more precise and meaningful identification. By increasing the number of features that can be taken into account in classifying organisms it enables the resulting classification to provide a more stable and useful framework to which all biological knowledge can be related. By demonstrating the detailed structure of organs that play an important part in the behaviour and life cycle of a species, such as the sense organs of insects, it enhances our understanding of the whole way of life of the species. Such understanding may be of immediate value in enabling us to control or conserve the species concerned — it also helps us to see ourselves in perspective as only one of the million and a half different species of organisms that share the resources of the earth, each one unique in its structure and way of life, and even the simplest having its component atoms and molecules arranged in a pattern that is mind-boggling in its complexity.

The photographs in this book were all taken or selected by Mr D Claugher in the Electron Microscope Unit of the British Museum (Natural History). Many other members of the Museum's staff have contributed — staff of the Photographic Unit by producing the prints, and staff of all the Museum's scientific departments in providing specimens and information and contributing photographs in their specialized fields.

*G B Corbet*

# How a scanning electron microscope works

The performance of a conventional, optical microscope is limited by the wavelength of its source of illumination. High-energy electrons have effective wavelengths that are about a thousandth of those of light and this was the main reason for the development of the electron microscope.

The scanning electron microscope (SEM) displays a magnified image of a specimen surface on the screen of a cathode ray tube (the 'picture tube' of a television set). Modern SEMS have a magnification range from x10 to x100000 and a resolution of three nanometres can be achieved, i.e. the ability to detect separately two points three nanometres apart (a nanometre being a millionth of a millimetre). Even at low magnifications a better image can be produced than with the light microscope due to the depth of field, which can be some 300 times greater than is possible with glass lenses.

There are some disadvantages however. The path taken by the electrons must be maintained at a high vacuum so that they are not displaced from their intended trajectory by collisions with gas molecules. This makes live specimens difficult to examine and soft tissue must be pre-treated so as to be stable at low pressure. For the best resolution most specimen surfaces must be made electrically conductive by depositing on them a very thin layer of metal — usually gold or a gold alloy.

The ability of the SEM to produce an image depends upon the similarity of its electron optical column to a cathode ray tube. They both have a 'gun' that fires a beam of high energy electrons through a series of magnetic fields designed to focus the beam to a fine spot.Both employ other, varying fields to deflect the focused beam to form a rectangular pattern of lines similar to that produced on the screen of a domestic television set. This pattern is called a raster and, in the cathode ray tube, may be used to create a picture by varying the intensity of the spot as it travels over the tube face.

In the electron optical column the electrons are generated by an incandescent tungsten wire filament and accelerated to form a diverging beam by an electric field of between 1000 and 40000 volts. A series of two or three magnetic lenses then focus the beam, producing successively smaller images of the source until it is brought to a final focus as a tiny spot on the surface of the specimen. (Magnetic lenses are so called because the magnetic field concentrated at their centre has a similar effect on the electron beam to that of a glass lens on a beam of light.)

A scanning electron microscope in action. The vertical
vacuum column is at the left, with the electron gun at
the top and the controls operating the specimen
chamber at the bottom.

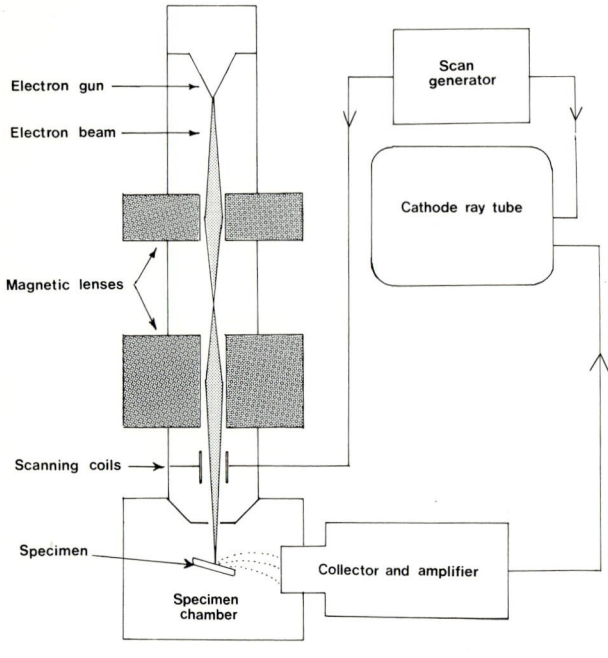

Electron gun

Electron beam

Magnetic lenses

Scanning coils

Specimen

Specimen chamber

Scan generator

Cathode ray tube

Collector and amplifier

Electron optical column

A diagrammatic representation of the working of a scanning electron microscope.

Voltage waveforms are applied to the scanning coils of both the column and the cathode ray tube by the scan generator. The resulting fields deflect the electron beam in the cathode ray tube so that it 'draws' a visible raster on the phosphor-coated tube face. The same waveforms, greatly reduced in size, are simultaneously fed to the column scanning coils causing a very small raster to be 'drawn' on the specimen surface. The ratio between the sizes of these rasters determines the magnification of the microscope image and may be varied by the operator.

When the high-energy electron beam strikes the specimen, secondary electrons are emitted, their number and energy being dependent upon the shape and nature of the specimen at that point. These secondary electrons are collected and amplified to form a fluctuating signal that is used to control the brightness of the cathode ray tube beam. Its intensity at the tube face is thus varied in precise synchronism with the varying emission of electrons from the specimen surface, creating a magnified image.

As well as secondary electrons, X-rays and reflected high energy electrons leave the specimen. Different types of collector allow these emissions to be used in the production of X-ray maps and in the analysis of the chemical elements present in the surface.

In studying vertebrates — mammals, birds, reptiles, amphibians and fishes — scanning electron microscopy has been used especially to show the detailed structure of the external hard parts, i.e. the skin and its derivatives and the teeth. Structures such as hair, feathers, scales and teeth have in common the fact that once the characteristic form is achieved during development the tissues die and there is therefore little further change in the structure other than by abrasion, although of course the dead tissue is often shed and replaced by new. This is in contrast to the internal hard structures — the bones — which are living tissues and are constantly being modified in response to the stresses of the moment.

# Human sperm

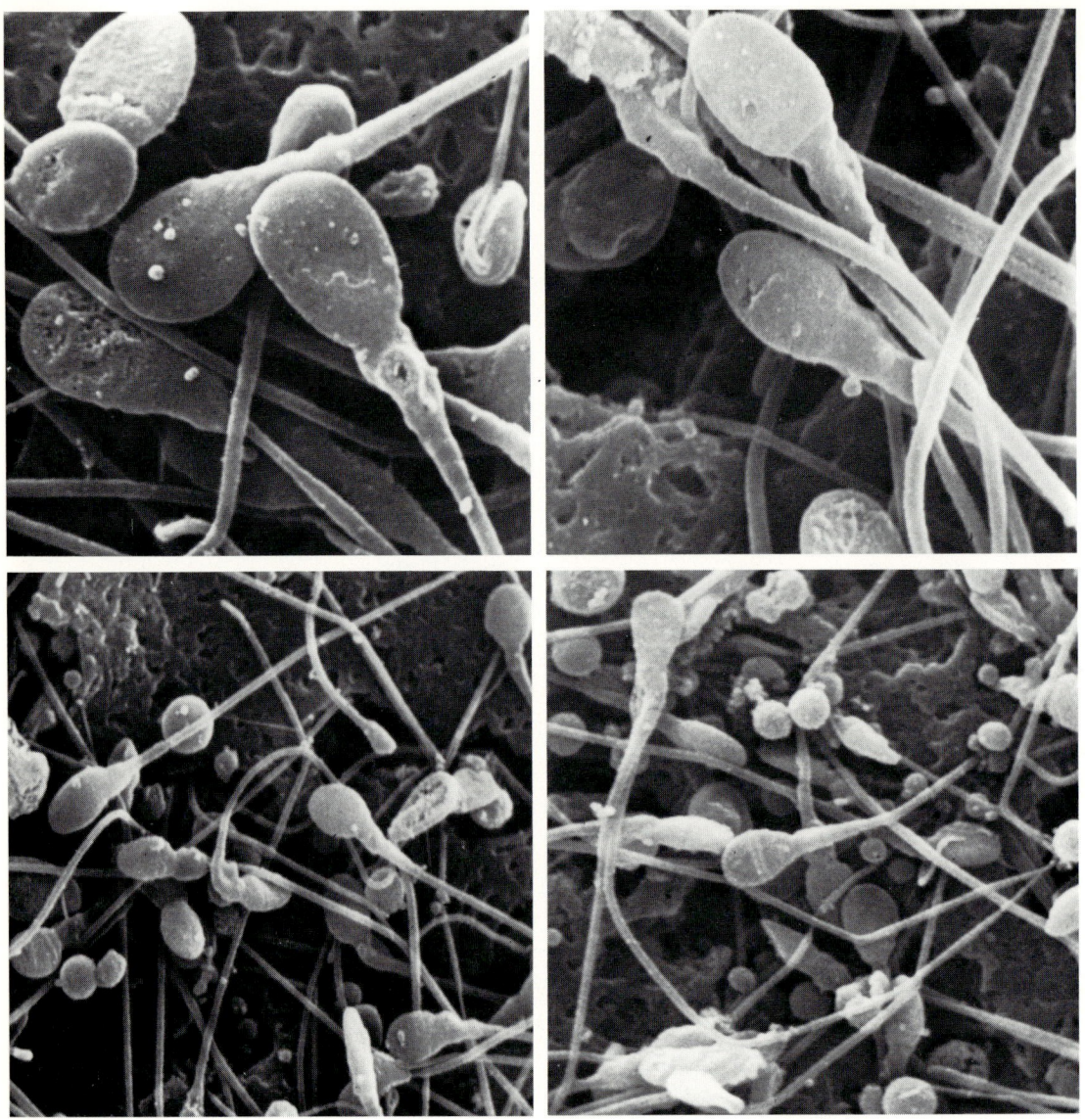

*Mature spermatozoa [above, x 5000; below, x 2140]*

Sperm cells or spermatozoa are amongst the smallest cells in the body, smaller, for example, than red blood corpuscles. The oval head contains the chromosomes (already reduced to half the number present in other cells of the body). The first, slightly thickened part of the tail provides the energy and the long filament serves for propulsion. A normal ejaculation contains about 500 million sperm cells. There is considerable variation in their shape in different species of animals.

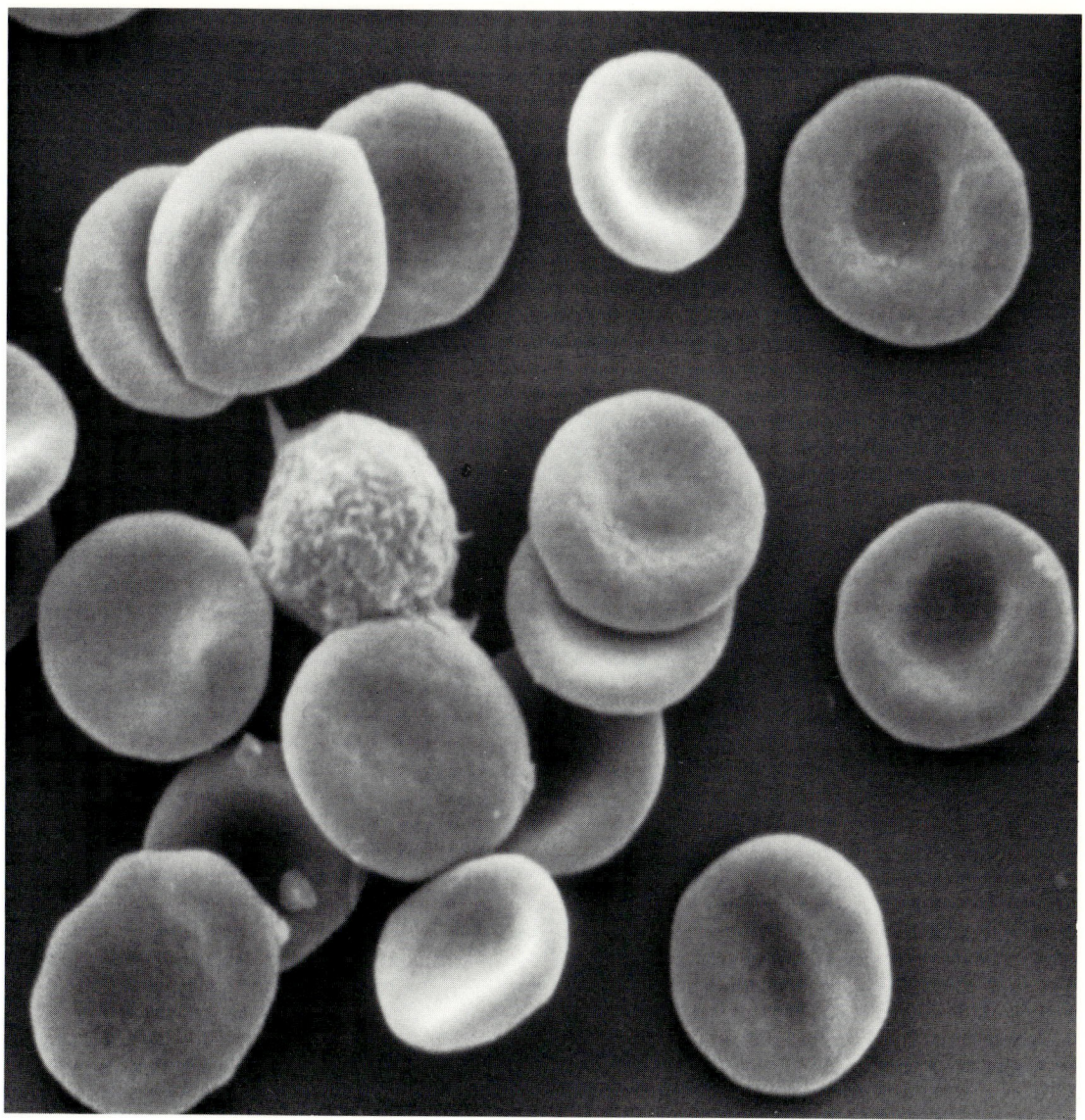

*Red blood corpuscles* [x 5350]

The red pigment haemoglobin, which carries oxygen to all parts of the body, is contained in the red blood corpuscles or erythrocytes. These outnumber other cells in the blood by 20 to 1, with a density of about five million per cubic millimetre of blood. They are unusual amongst the cells of the body in lacking a nucleus. They have a life of about 120 days between their production in the bone-marrow and their destruction, mainly in the spleen and liver.

# Human hair

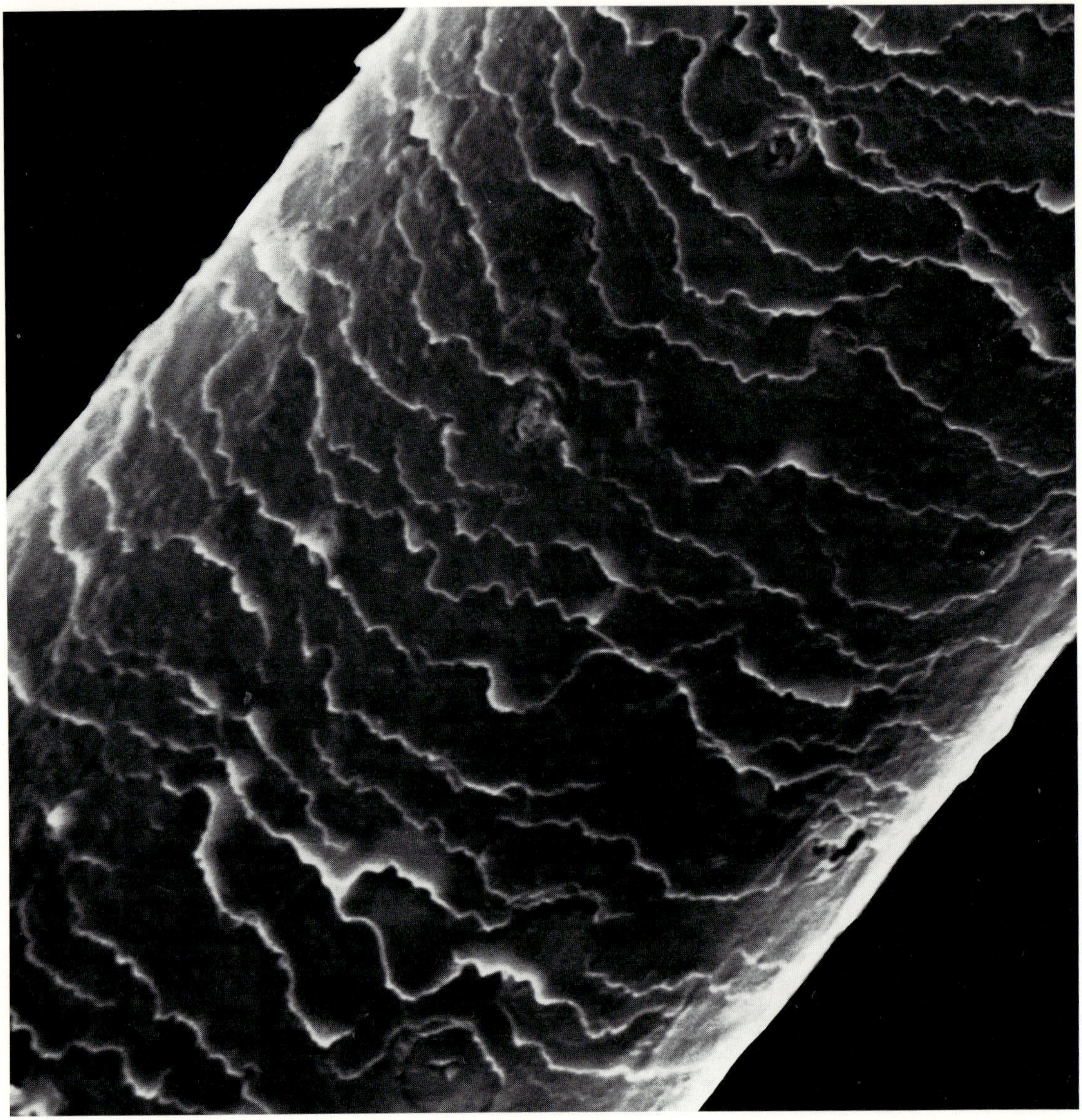

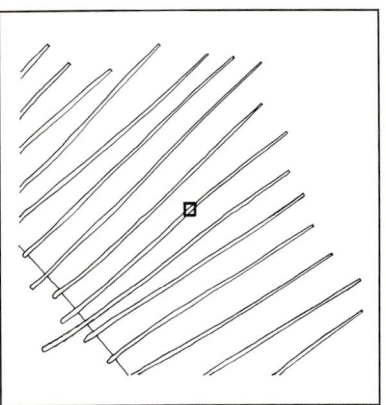

*Outer surface* [x 1600]

Hair is present on most parts of the body and wherever it is found the microscopical structure is much the same. The irregular pattern seen in this photograph is formed by layers of flattened dead cells that overlap each other. This scaly layer is known as the cuticular layer or cuticle.

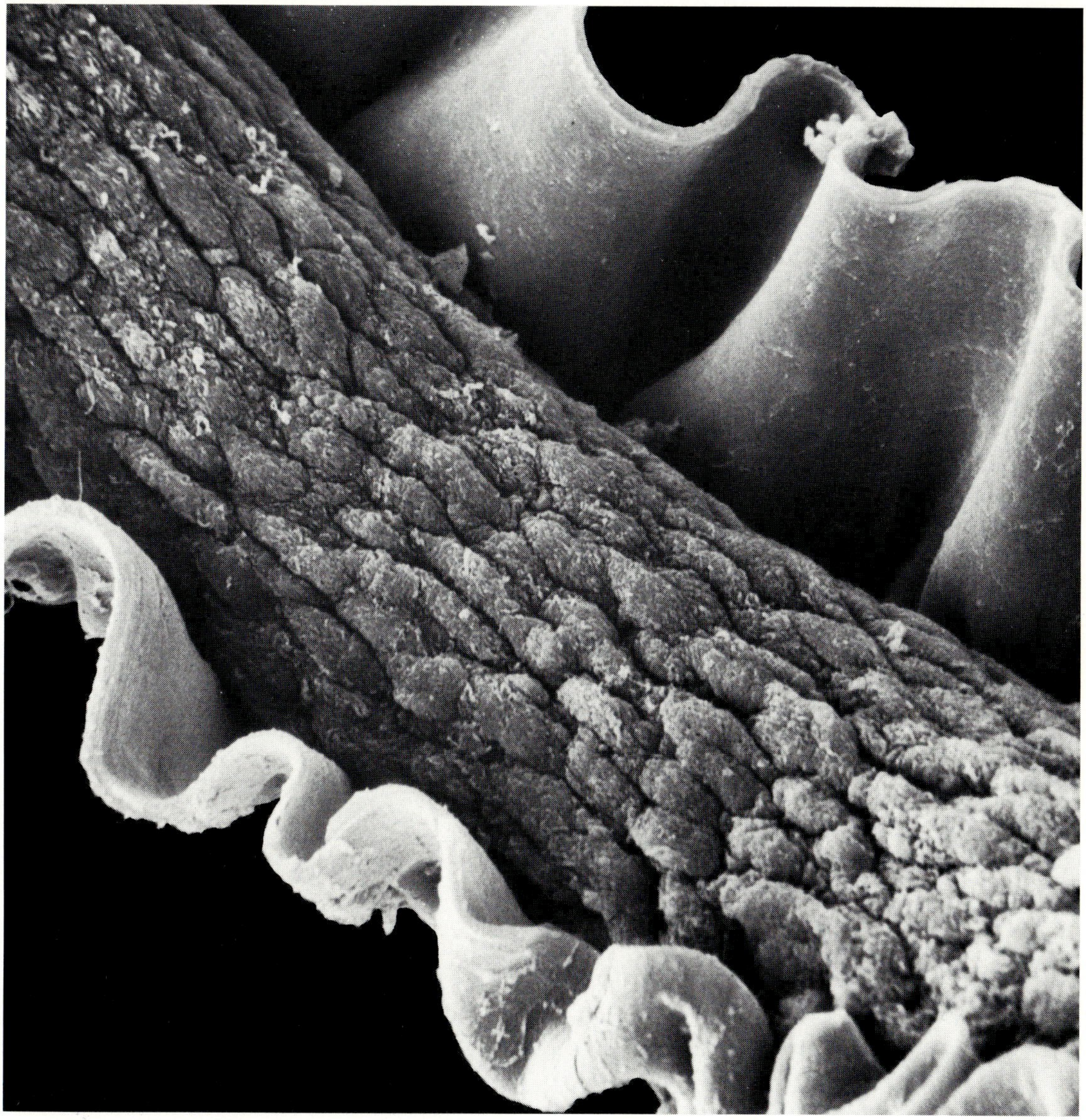

*Cuticle reflexed* [x 685]

This unusual photograph shows the cuticular layer of the hair cut open and reflexed to expose the core or cortex. The cortex is a very complex structure but a large part of it consists of the dead remnants of nuclei and cell membranes. It is rather soft in texture, depending upon the cuticle for support. It has no nerve nor blood supply.

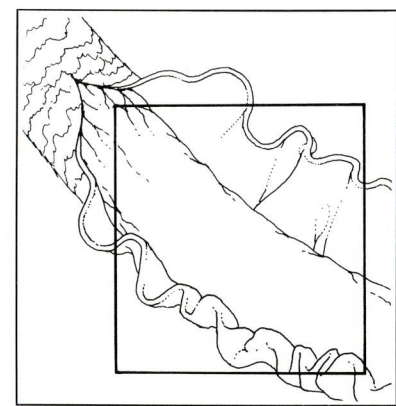

# Human hair

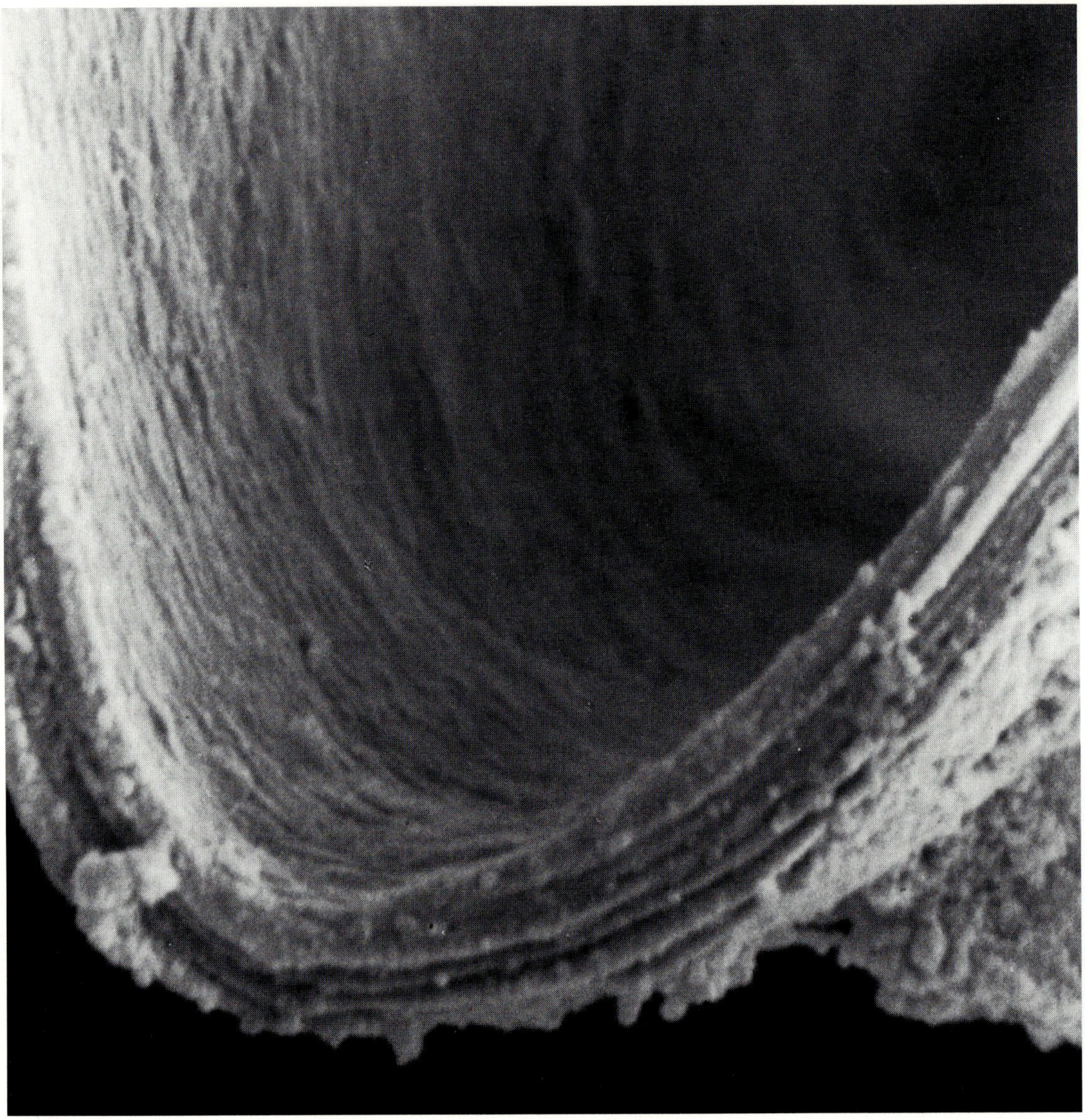

![Section of cuticle]

*Section of cuticle* [x 6900]

Part of the preceding specimen enlarged to show the layering visible in the cut surface of the scale-like cells that make up the cuticle. These cells are composed largely of the protein keratin, produced in the hair follicle in the skin as the hair is growing.

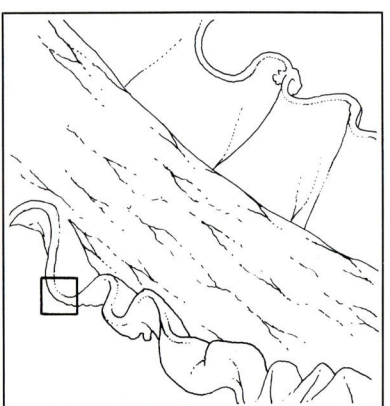

*Finger-tip* [x 105]

The outer covering of our bodies is composed of epidermal cells that have become keratinized. Keratin is a tough fibrous protein containing sulphur, which forms a resistant layer not only in the epidermis of the skin but also in hairs, scales and feathers. The skin figured here is from the tip of a finger and includes a number of sweat pores (surrounded by faint circular markings).

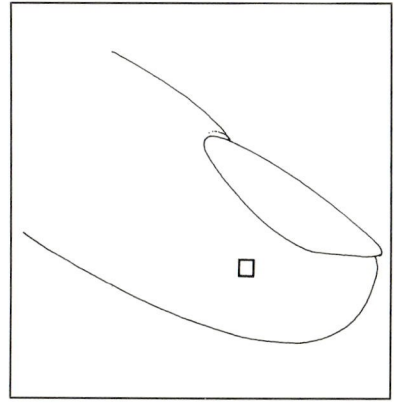

# Human tooth

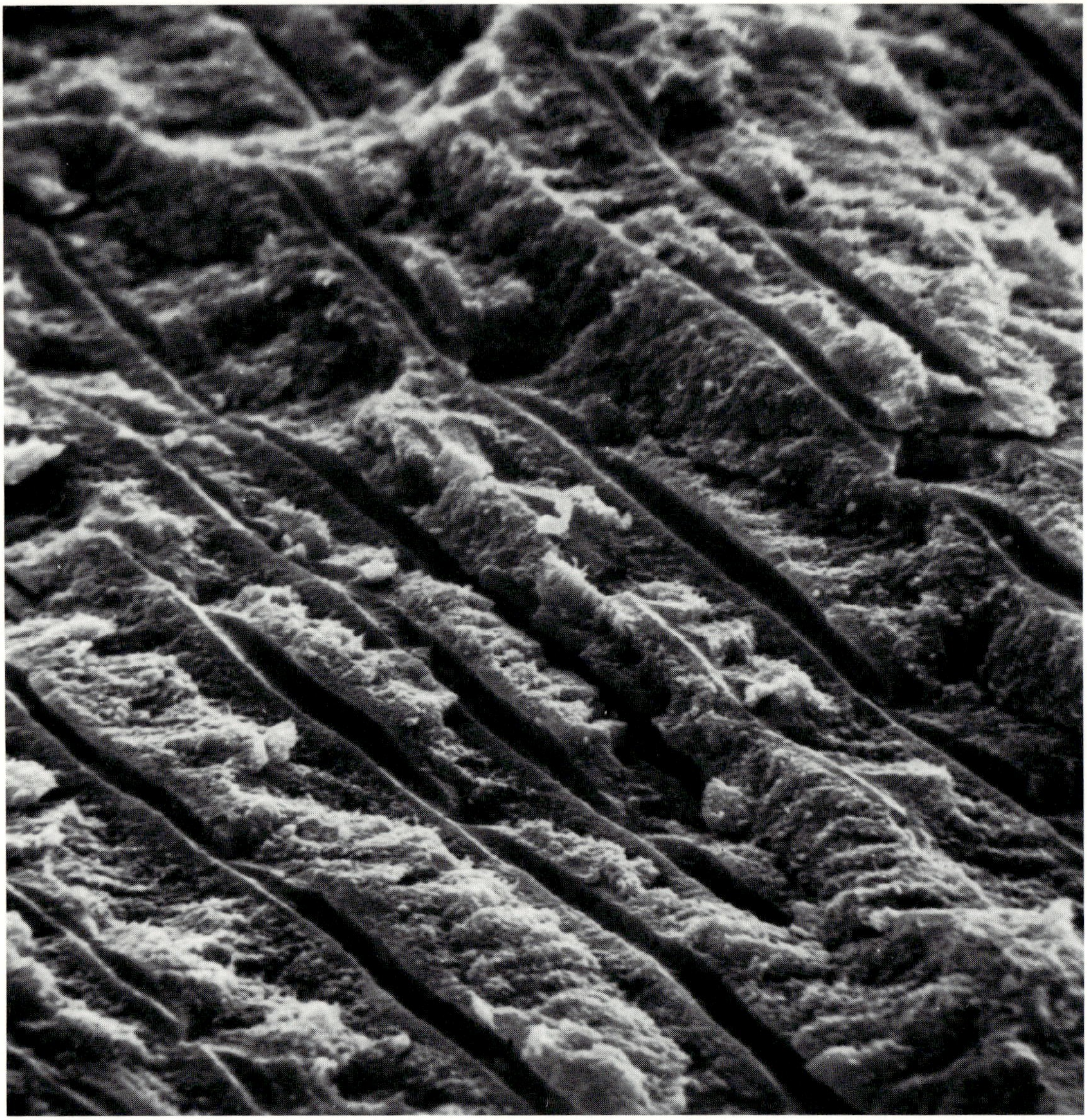

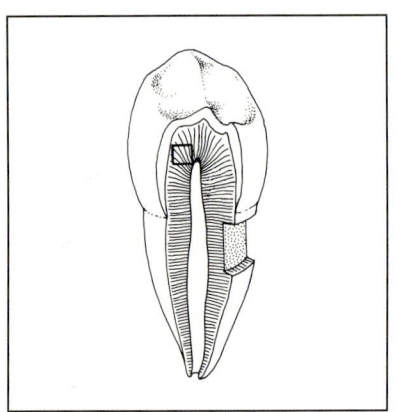

*Longitudinal section of tubules in dentine [x 2760]*

In most mammals the teeth are made up of four layers of material: a soft living core or pulp; a tough middle layer of dentine; a hard outer protective layer of enamel; and cement which lies between the dentine and the socket or contact area of the tooth and the jaw. This photograph shows the fine tubules or *Schreger's lines* that pass through the dentine. These tubules are 0.005 mm (i.e. 5 microns) in diameter and run between the pulp cavity and the enamel or cement.

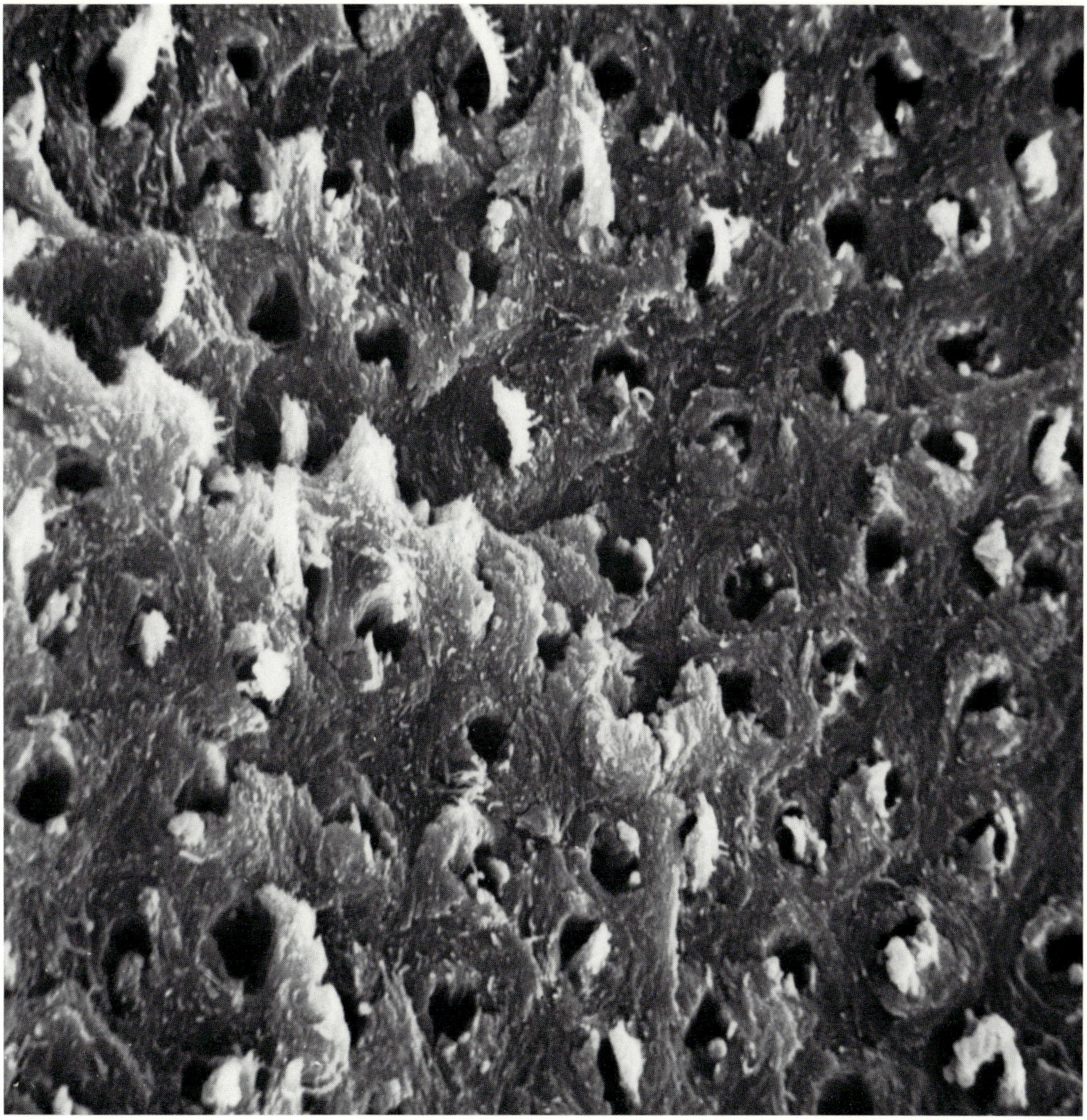

*Transverse sections of tubules in dentine* [x 2760]

A fractured surface of dentine. In life the tubules contain fluid as well as fine processes of the odontoblasts, the cells in the pulp cavity that produce the dentine. These are not visible in this photograph.

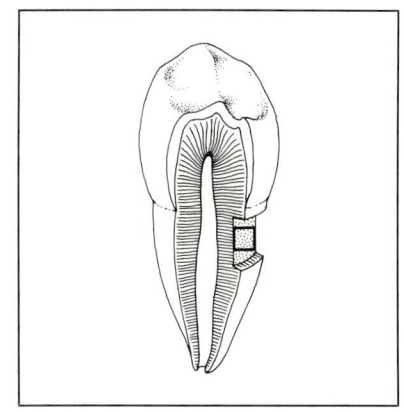

# Ivory

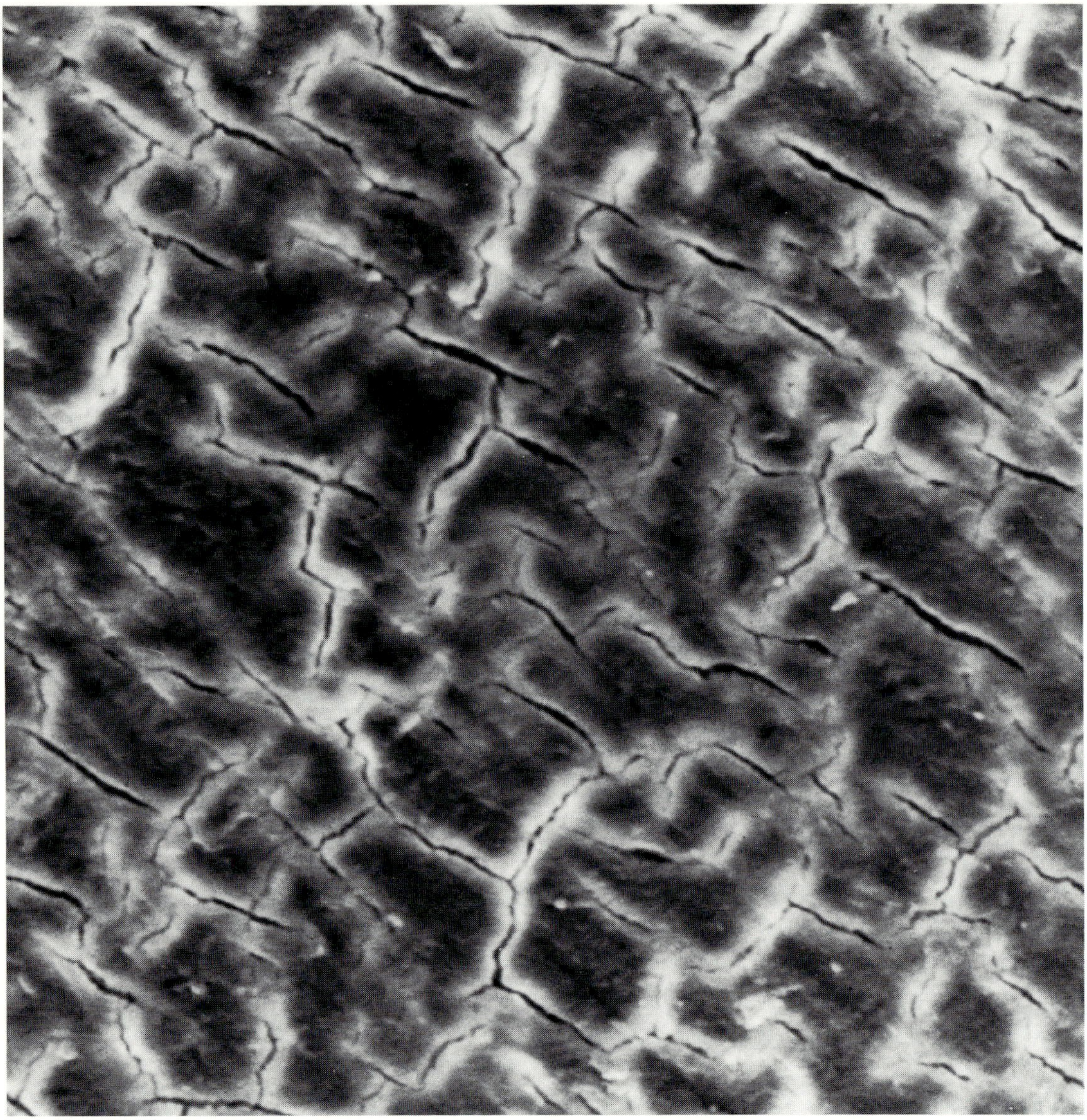

*Polished surface of ivory from an Indian elephant* [x 1380]

The tusks of an elephant represent the upper incisor teeth and are usually confined to the males in Indian elephants although both sexes bear tusks in the African species. By examining a polished surface it is possible to distinguish between elephant ivory and that from other animals like walrus and sperm whale, but there is no satisfactory way of differentiating between the various species of elephant, including mammoth.

# Egg of domestic hen

*Inner surface of shell membrane* [x 360]
The liquid content of the egg is contained by a membrane composed of fine intertwined fibres. This photograph shows a thin layer, probably of dried albumen (egg-white), with small cracks through which the fibres of the membrane can be seen.

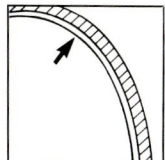

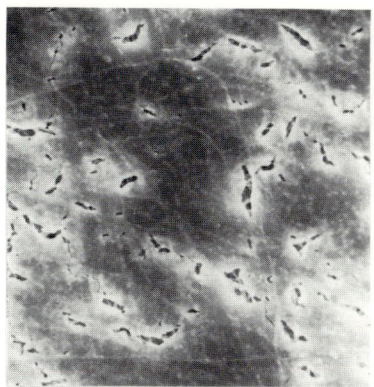

*Outer surface of shell membrane* [x 730]
This fibrous membrane not only contains the albumen but is thought to protect the egg from penetration by micro-organisms that can pass through pores in the hard shell.

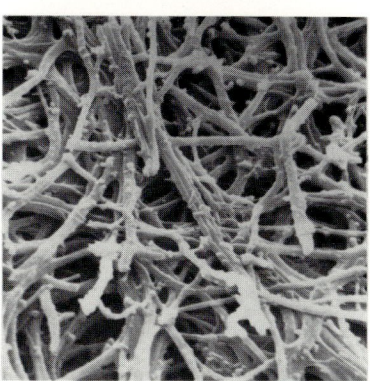

*Inner surface of the hard shell* [x 145]
The shell is formed on the outer surface of the soft membrane by the growth of crystals from a large number of separate points. These coalesce as they grow outwards but leave a number of fine canals penetrating the shell.

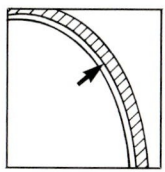

*Outer surface of the shell* [x 70]
The irregular pattern of fissures varies from one group of birds to another. The openings of the canals can be seen as very small, rounded pores. These allow oxygen to pass to the developing chick and waste carbon dioxide to escape.

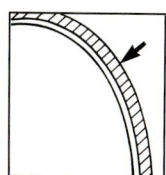

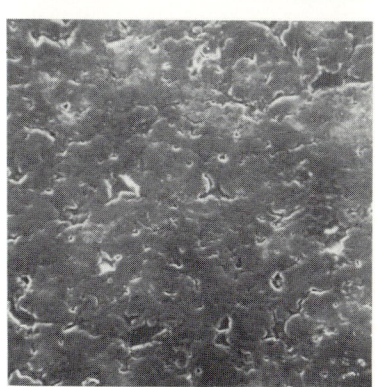

# Feather

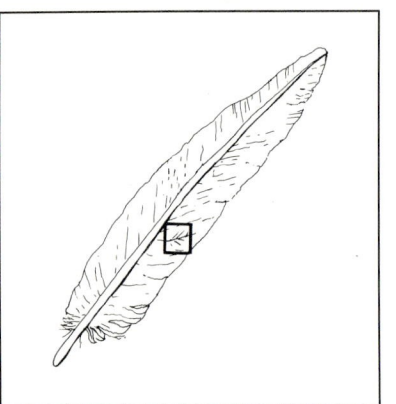

*Surface of vane showing parts of eleven barbs [x 65]*

The barbs of a feather are the parallel branches on either side of the central shaft. Each barb carries smaller parallel branches (barbules) on both sides. Those on the distal side of the barb (pointing towards the tip of the feather) carry hooks (pale areas in photograph) which interlock with the smooth barbules on the adjacent (proximal) side of the next barb.

*Parts of three detached barbs* [x 20]
The barbs have been separated to contrast the distal, hooked barbules (the frilly edges on the right) with the proximal, smooth barbules on the left.

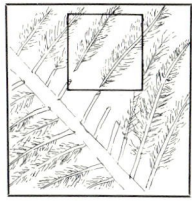

*Part of a single barb* [x 1100]
The articulation of four barbules can be seen.

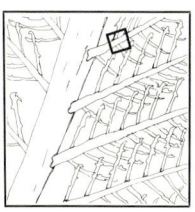

*Two interlocked barbs* [x 100]
The hooks on the distal barbules can be seen overlapping the smooth proximal barbules.

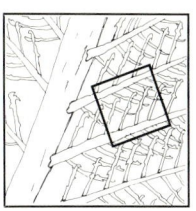

*Detail of barbules* [x 1000]
The hooks of the pale, distal barbules can be seen gripping the smooth dark proximal barbules.

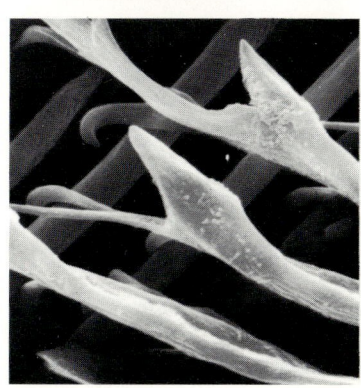

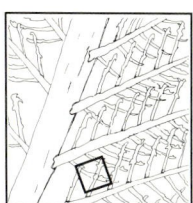

# Dogfish

*Skin of Lesser spotted dogfish* [x 65]

The Lesser spotted dogfish, *Scyliorhinus canicula,* is a small shark common in British waters. Like other cartilaginous fishes (elasmobranchs) its body is covered with small dermal denticles or placoid scales that are quite unlike the flat round scales of bony fish (teleosts) such as herring or cod.

# Dogfish

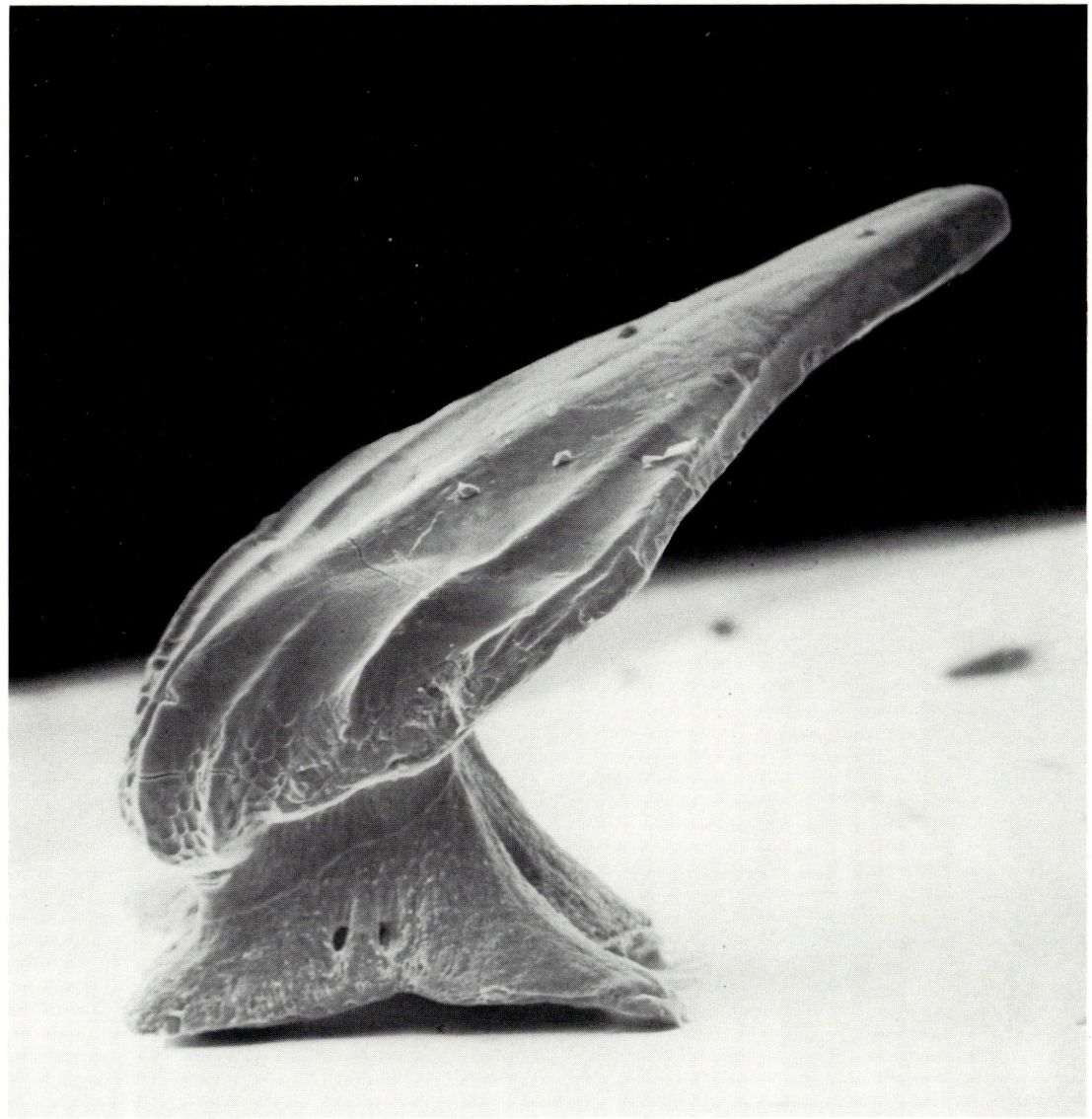

*Side view of a single placoid scale [x 140]*

A placoid scale resembles a tooth. It has a bone-like base of dentine with a pulp cavity embedded in the skin. Projecting from this base is an arrow-shaped spine of dentine sheathed with enamel, as in a tooth. Hence shark scales are often called denticles, literally 'little teeth'.

# Wool

Sheep have been bred for wool by selecting for an abundance of the fine, dense underfur at the expense of the long, straight 'contour hairs' that form the sleek outer surface of the coat in most mammals. However, the wool fibres themselves vary greatly in thickness, especially in some mountain breeds such as the Scottish blackface, and this helps to trap air and contribute to the insulating property of wool.

Insects constitute the most numerous and diverse group of animals, with about 800 000 different species already described and many more being discovered each year. In spite of this diversity, they have many features in common, including a tough and flexible cuticle containing chitin, a substance that is chemically related to cellulose. Chitin is hardened to make an external skeleton by the deposition of sclerotin, formed from a tanned protein. This serves both for protection and as a firm attachment for the muscles. This outer integument is, like the outer layer of vertebrate skin, dead material, but it is shed as a complete skin at intervals during growth of the insect. The structure of this exoskeleton, which covers the entire body, therefore remains constant during each stage or instar between moults and shows a remarkable range of detail in texture, shape and protuberances.

# Aphid

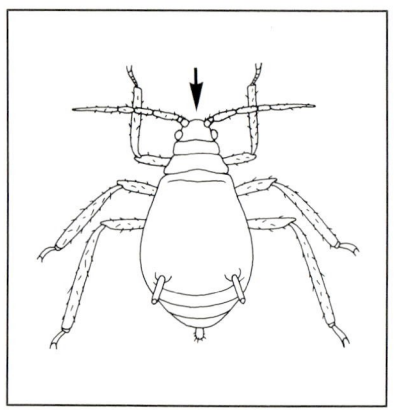

*Frontal view of black bean aphid,* Aphis fabae, *on a leaf* [x 160]

Aphids are insects that feed by probing the tissues of plants with a beak carrying long slender stylets. These form a fine tube through which the sap is sucked up like milk through a straw.

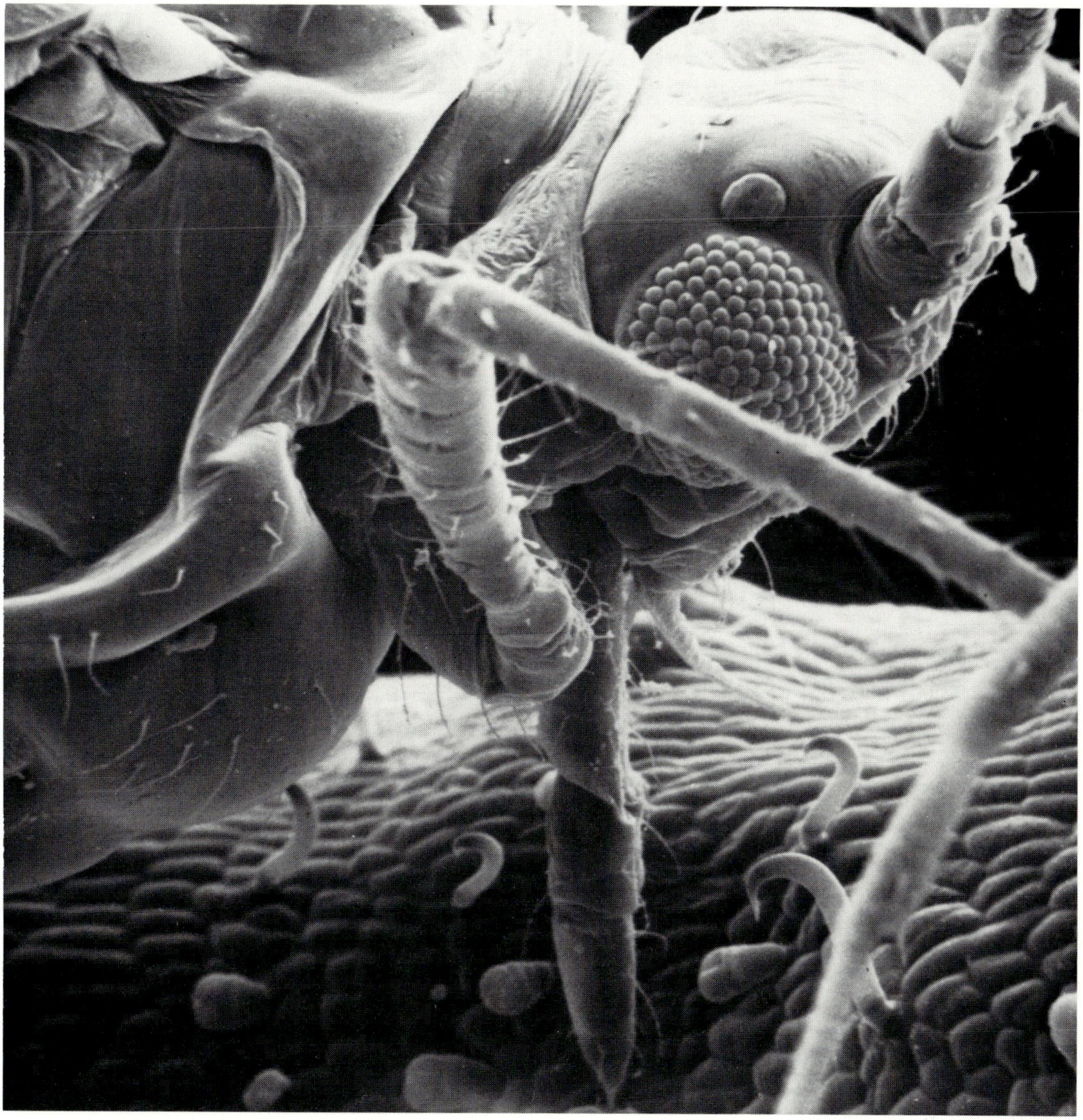

*Side view of head and thorax* [x 220]

Each side of the head is dominated by a large compound eye. The small circular structure above the compound eye is a simple eye (ocellus) characteristic of the winged adult form.

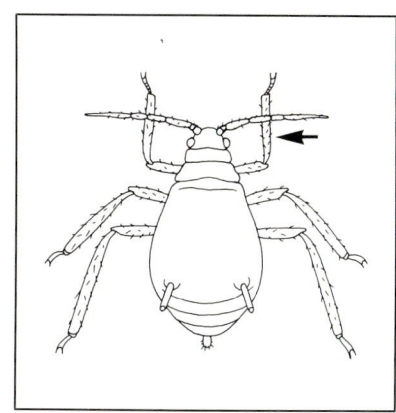

# Aphid

*First-stage young on leaf* [x 135]

Young aphids resemble adults in general appearance but are never winged. The small size can be appreciated in relation to the hooked hairs on the surface of the leaf.

*Eggs on a leaf* [x 150]

These eggs are of a moth of the family Notodontidae.
The circular, lid-like structure is unusual in moth eggs
and probably facilitates hatching of the larvae. The
central rosette-like structure is the micropyle
(meaning 'little door') through which the sperm enters
during fertilization and which also acts as a breathing
pore during the development of the embryo.

# Ant

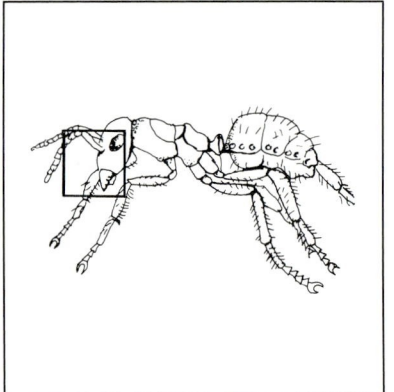

*Head of black garden ant*, Lasius niger [x 138]

Frontal view of the head showing the tip of the left antenna (top right), the left antennal socket (top centre), the toothed jaws and the ancillary mouthparts. This common garden ant has mouthparts adapted for lapping and it feeds mainly on sugary substances; the jaws are used as weapons, for manipulating aphids and other purposes.

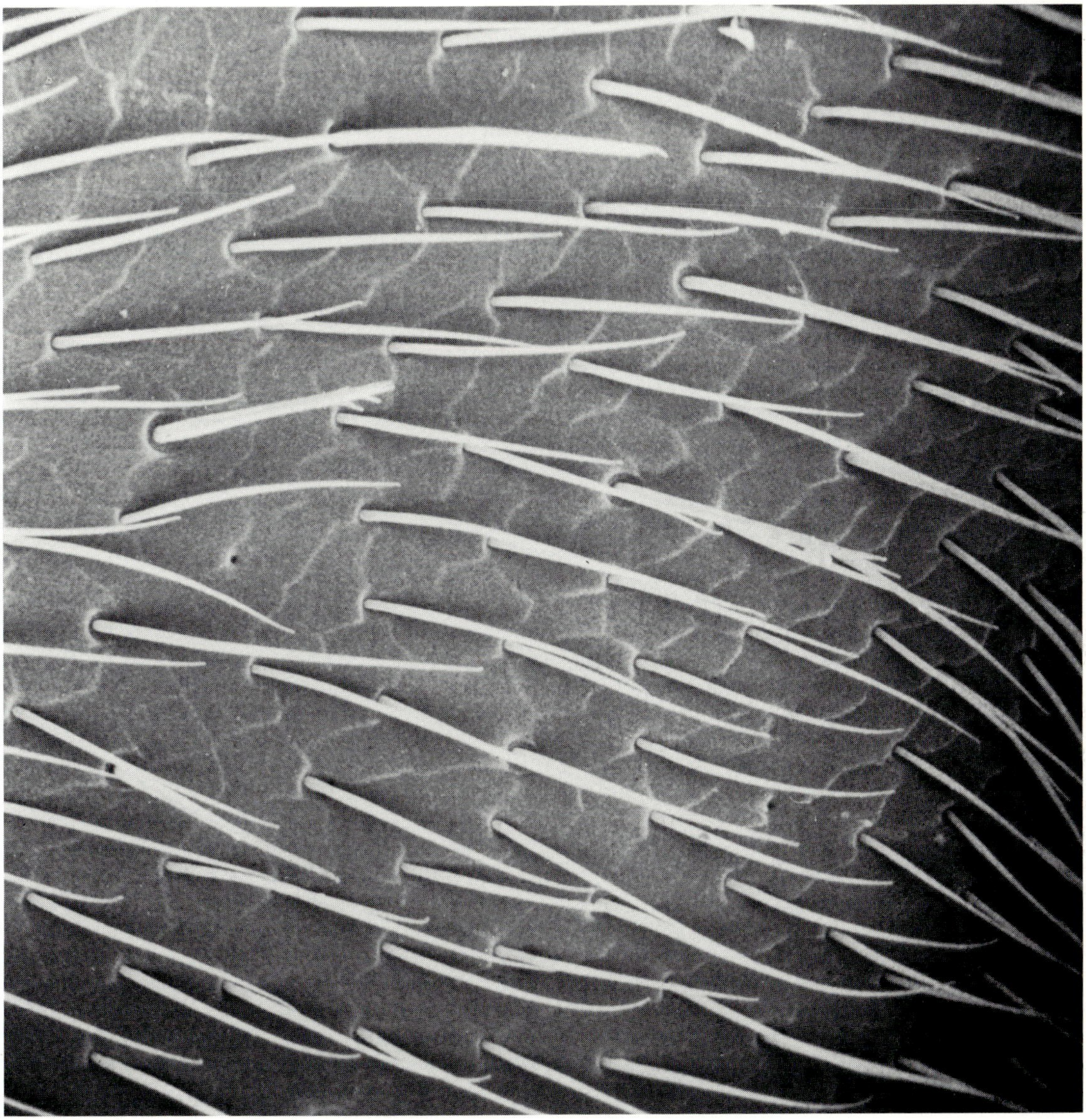

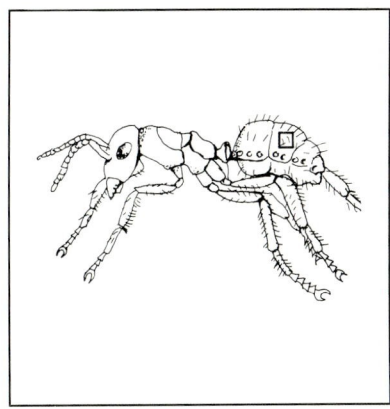

*Setae on abdomen* [x 690]

Insects are covered by a tough insensitive cuticle.
In order to feel, touch and smell they have to use
ancillary structures, the most common of these being
bristles called setae. Setae found on insects can be
divided into various groups depending on their
structure. The setae on the abdomen of an ant may be
mechanical touch receptors; each seta is set in an
individual pit and is attached to a nerve.

# Ant

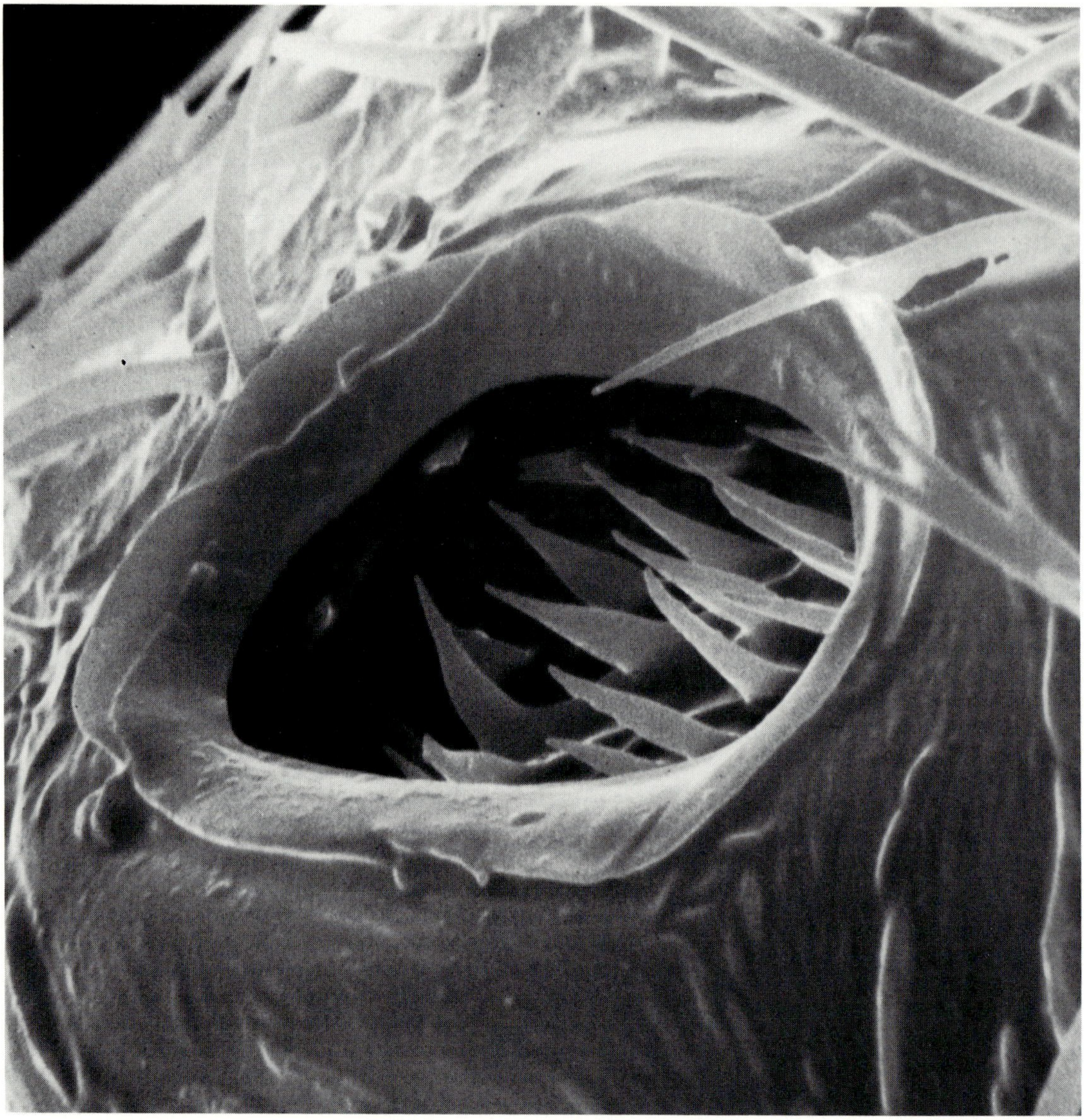

![SEM photograph of an ant spiracle]

*Spiracle* [x 2760]

In common with other insects ants breathe through a system of tubes that ramify throughout the body and are connected to the outside through openings know as spiracles. The triangular structures seen in this spiracle are sensory — if an ant encounters adverse atmospheric conditions it is capable of closing down its respiratory system and staying dormant for up to twenty-four hours.

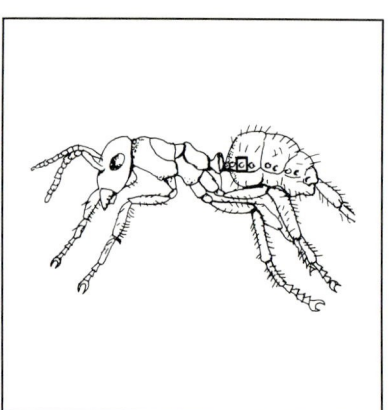

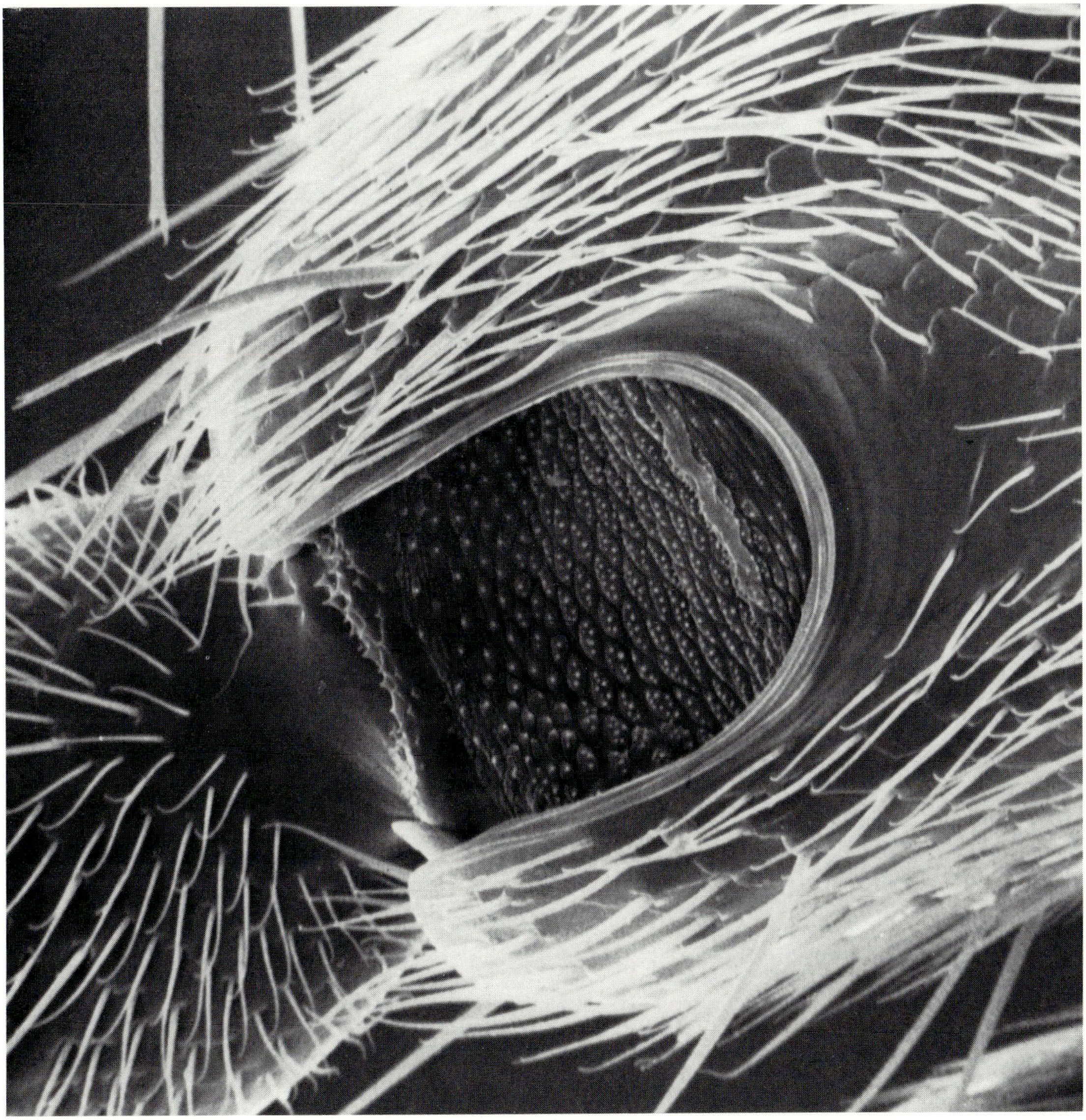

*Leg joint* [x 695]

The cuticle of most insects is tough and rigid, forming what is in fact an external skeleton. This poses problems where certain areas of the body have to be capable of flexion as in the abdomen and limb joints. This illustration of a joint on the leg of an ant shows the flexible area on the inside which is quite distinct from the harder parts of the cuticle. Towards the bottom left of the flexible area there are a number of rows of small spines in association with two pegs; this is the mechanism the ant uses to determine the exact position of the limb.

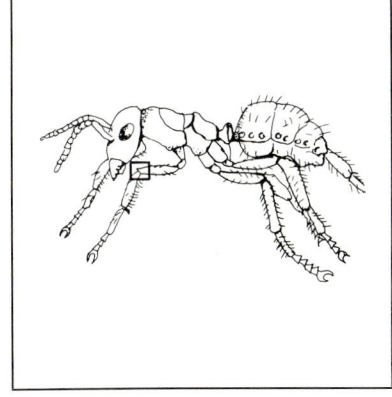

# Ant

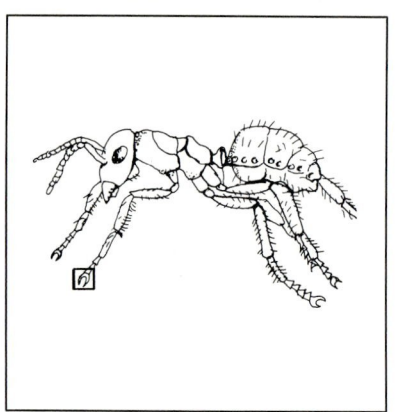

*Foot* [x 500]

The large setae are sensory and the two large claws are used for grasping. The circular structure between the claws is a suctorial disk which enables the ant to walk on the underside of smooth surfaces such as glossy leaves.

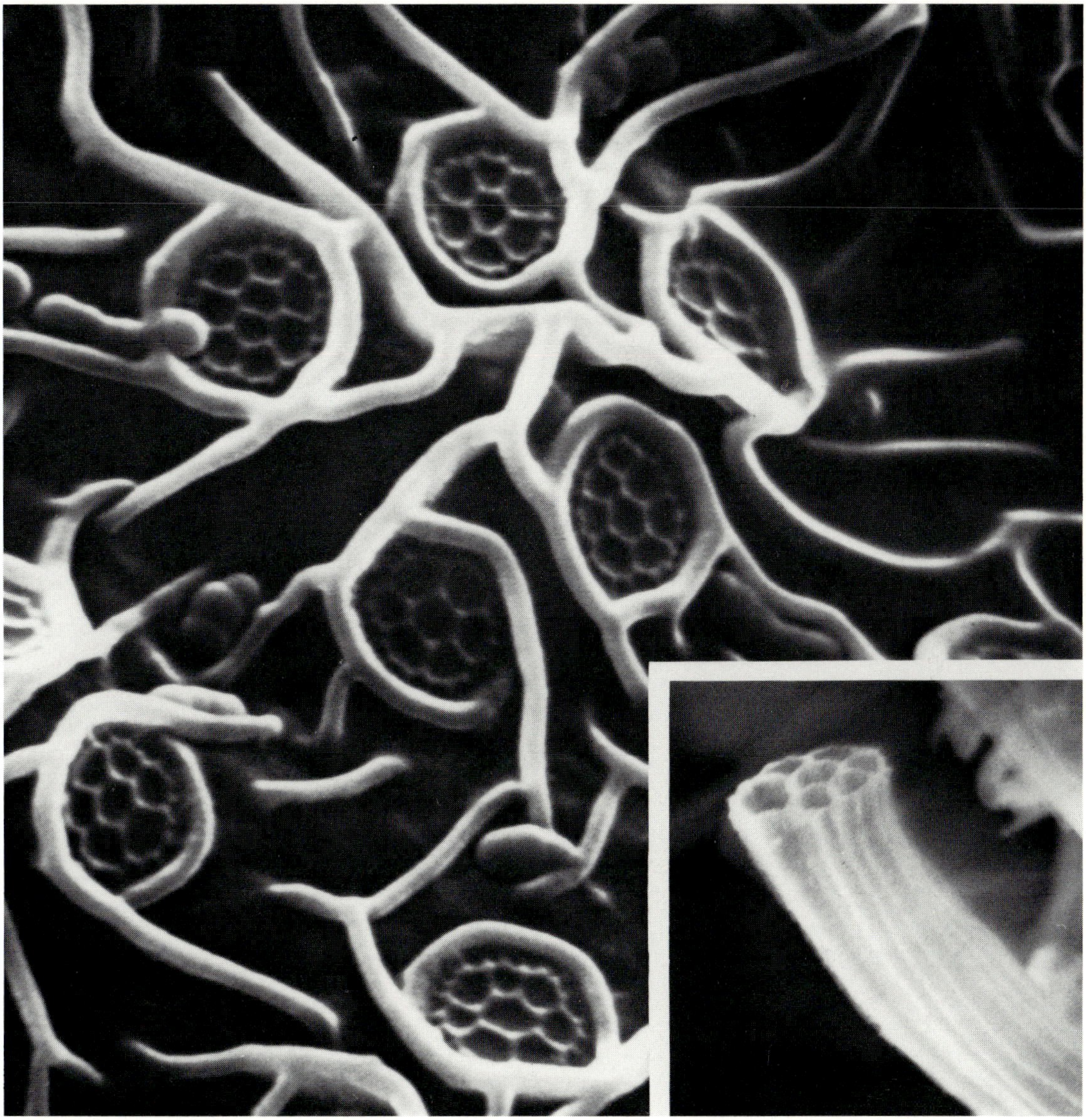

*Wax extruders on larva of a mealybug destroyer*
[x 7650]

This Australian ladybird, *Cryptolaemus montrouzieri*, has been introduced into various countries as a means of controlling mealybugs infesting citrus crops. The larvae disguise themselves by producing a cotton-wool-like coating of waxy material over their backs. The illustration shows some of the structures through which the wax (inset) is extruded.

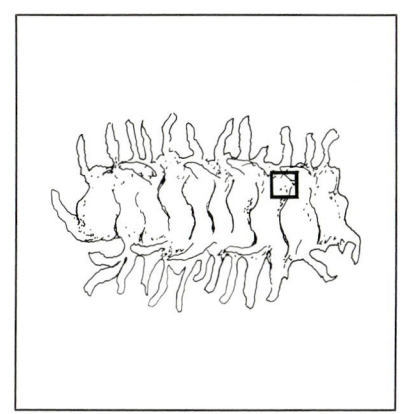

# Bumble bee

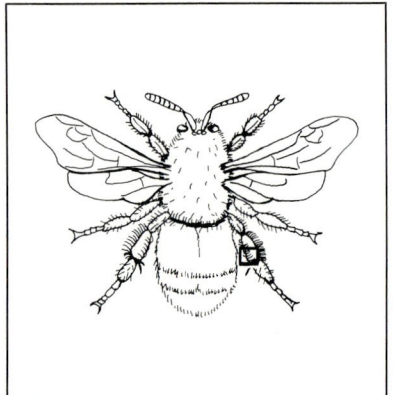

*Setae on the hind limb of a red-tailed bumble bee,* Bombus lapidarius [x 270]

The bristles or setae of insects are mainly sensory but some are adapted for other purposes. Those illustrated trap pollen which is subsequently combed off and deposited in the pollen basket on the hind legs.

# Bumble bee

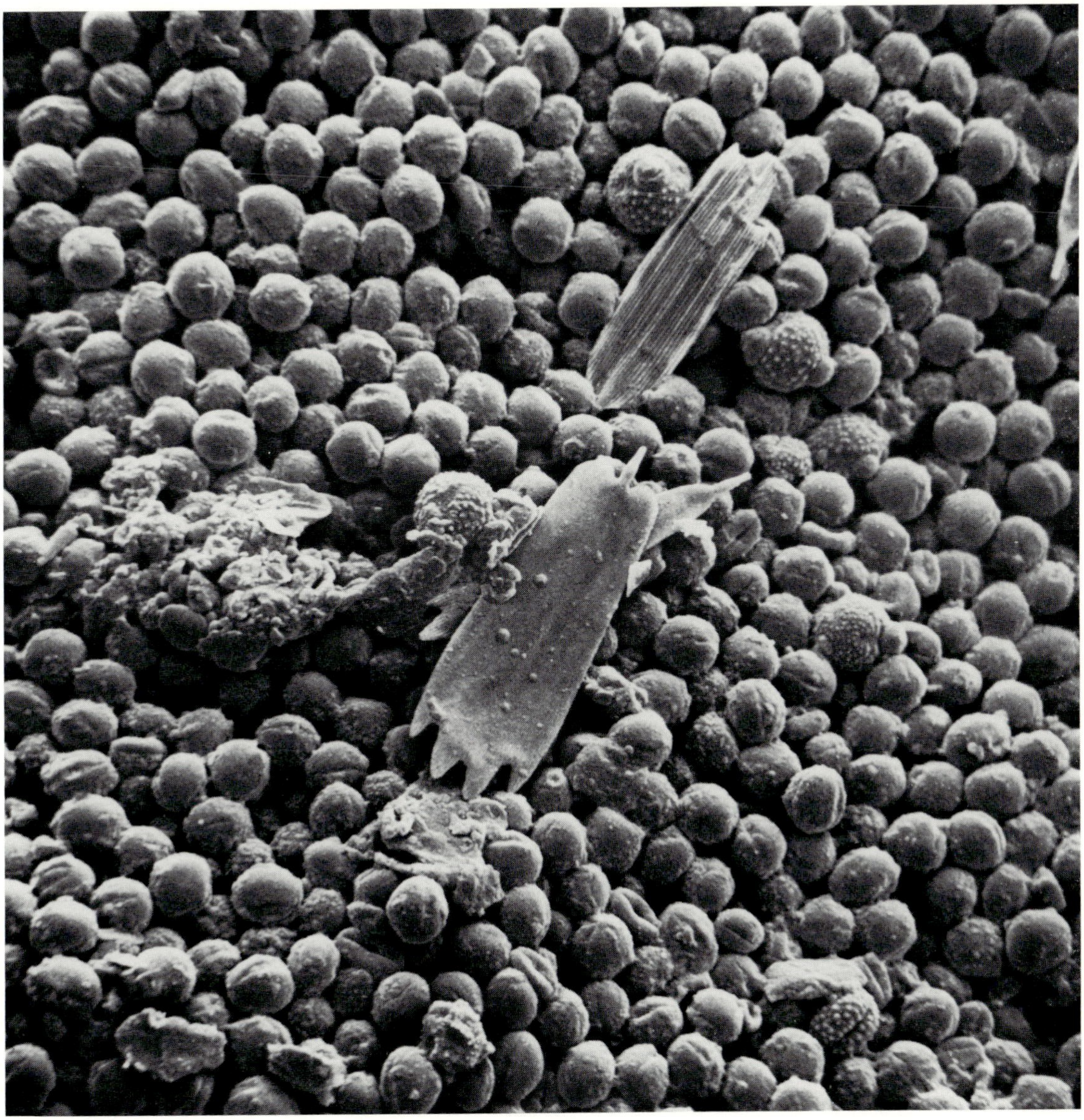

*Pollen load on the hind limb* [x 270]

Pollen collected by bees when visiting flowers often
sticks to the abdomen and other parts of the body.
The bee has a number of specialized hairs on the two
pairs of fore legs that pass the pollen to the mouth
where it is mixed with a little honey to make it sticky.
It is then passed back to the hind legs where it is
deposited in a specially developed 'pollen basket' for
transportation back to the nest. At this magnification
it can be seen that different kinds of pollen are present.
The larger structures are scales from a moth or
butterfly that may have brushed off on the flower and
stuck to the bee.

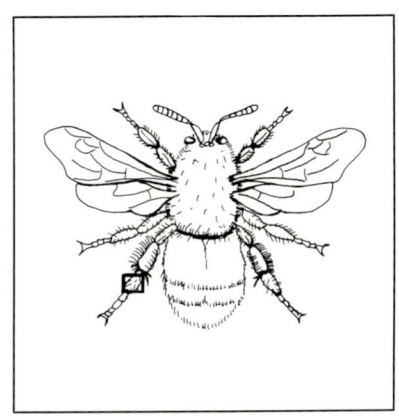

# Bumble bee

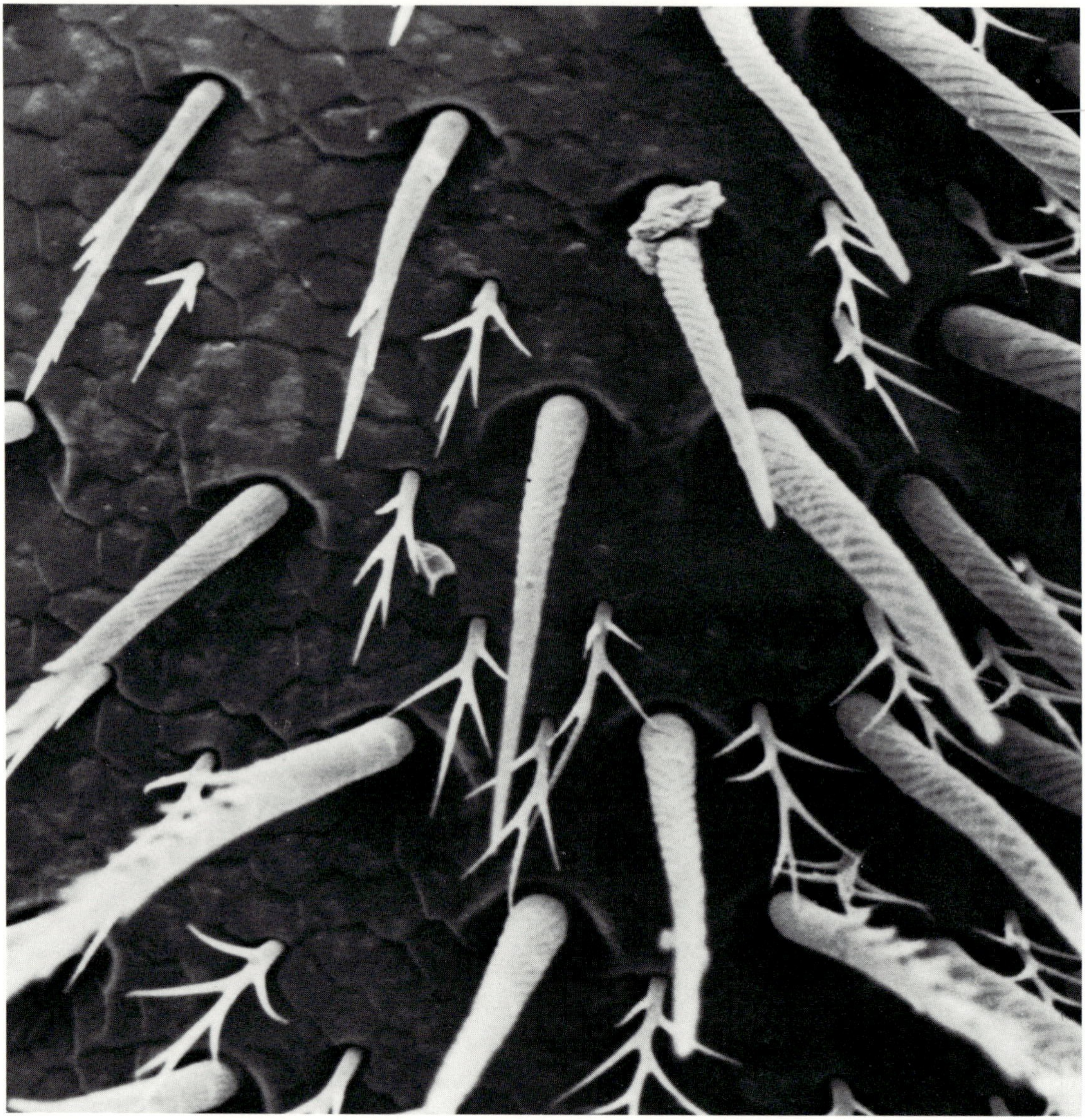

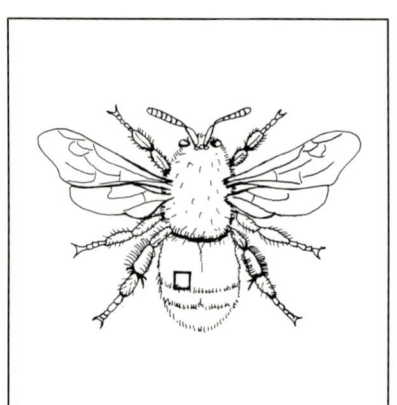

*Setae from the abdomen* [x 690]

There are a great variety of shapes and sizes of setae on insects but at present little is known of their function. The two kinds illustrated here are found on the abdomen and they probably have different functions.

*Comb on the fore leg* [x 140]

This structure is used to clean the bee's antennae.
The fore legs of bees and wasps have a fixed groove in
the first tarsal segment (upper left) close to the
articulation with the tibia. The tibia bears a separate
articulated spur, which may be closed against the
notch. The fore leg is raised over the antenna which
is inserted in the notch, the mechanism is closed and
the antenna is drawn through. The antenna bears a
number of sensory organs and if pollen or other matter
is left adhering to these it impairs their function.

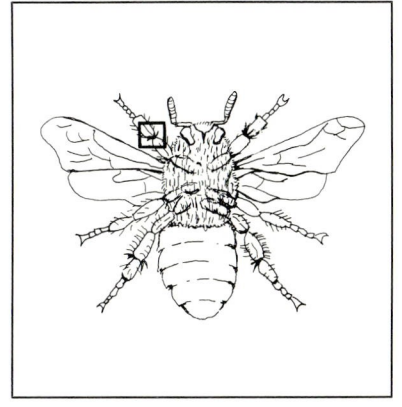

# Blowfly

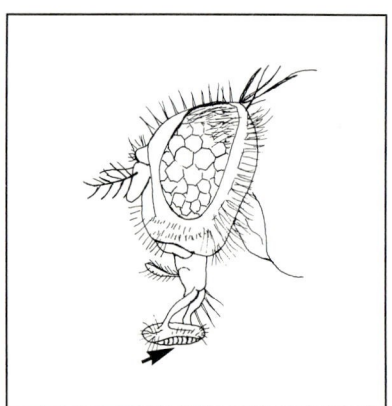

*Proboscis of a blowfly,* Calliphora vicina [x 140]

The mouthparts of most two-winged flies (Diptera) are modified for sucking up liquids, except in those species in which the adults do not feed when the mouthparts are atrophied. In predatory and parasitic species the mouthparts are modified both for piercing the prey or host, and for sucking up the body fluids. Blowflies and related groups exude salivary fluids, and partial digestion of the food takes place *in situ*, before the resulting fluids are sucked up by a pair of terminal pads on the mouthparts called labella. Each labellum bears specially strengthened channels on its inner surface called pseudotracheae.

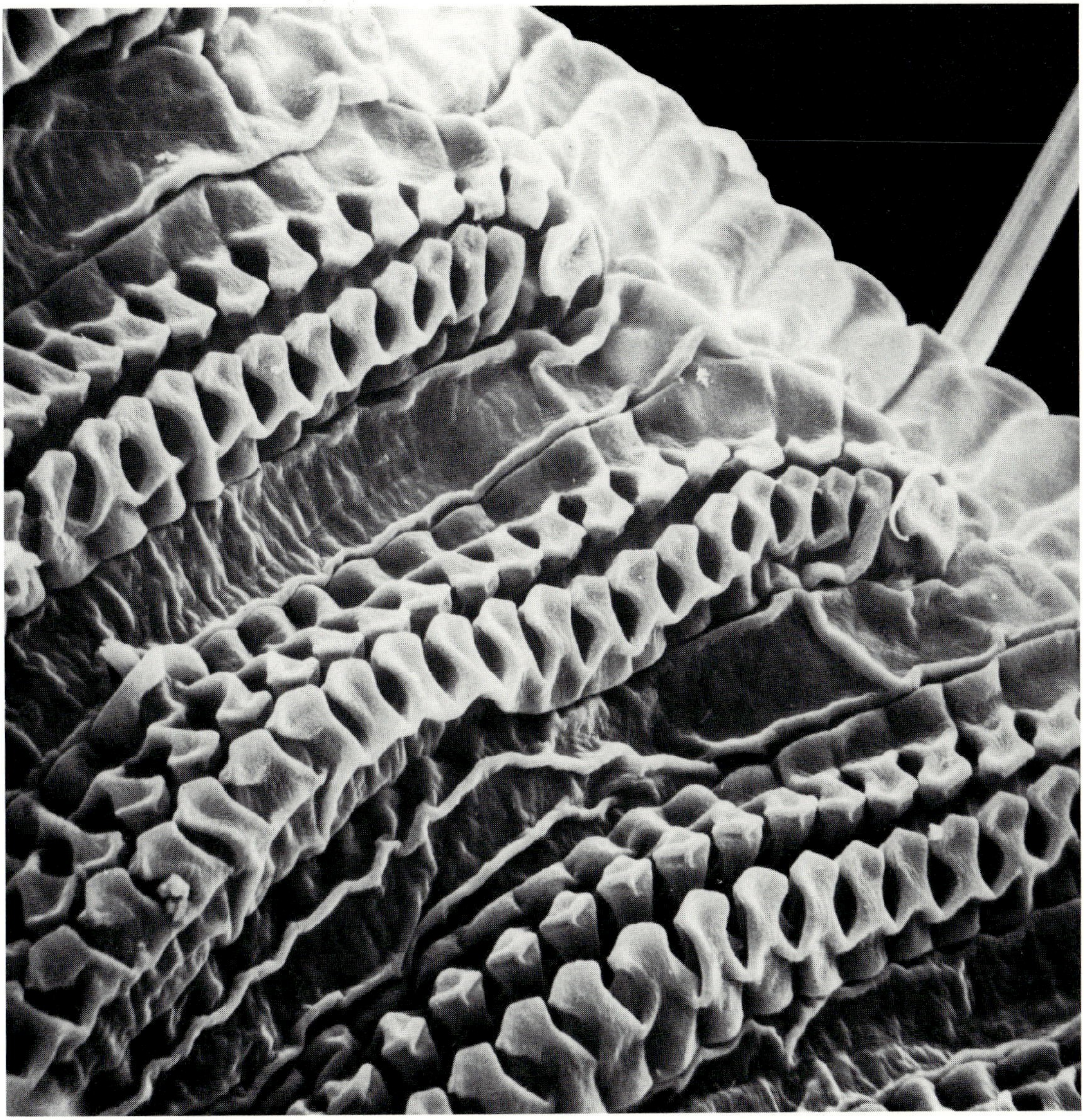

*Detail of proboscis* [x 1380]

An enlarged portion of the preceding picture, showing the pseudotracheae in greater detail. As part of their function the pseudotracheae act as filters, rejecting particles larger than 0.004 mm.

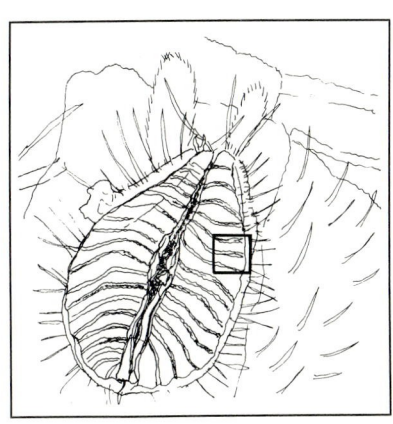

# Blowfly

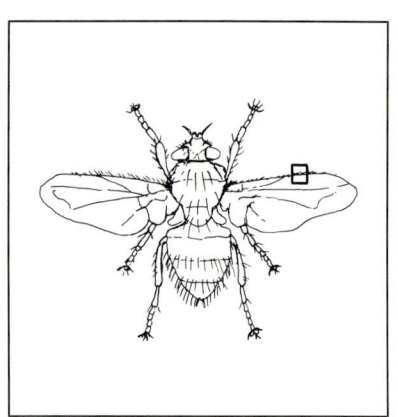

*Leading edge of wing* [x 955]

The spines and bristles shown may protect the leading edge of the wing from mechanical damage during flight, and may also have a sensory or aerodynamic function.

*Surface of wing* [x 275]

The small setae covering the wing surface are known as microtrichia; their function is not known, but they may serve to increase the aerodynamic efficiency of the wing by trapping a layer of air close to the wing membrane.

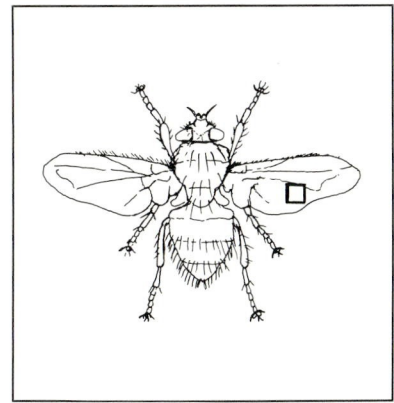

# Blowfly

*Foot* [x 275]

Note the two large claws that are common to many insects, and the various setae that project forward; the latter probably have a sensory function. The pair of large pads between the claws are covered with very small pads on slender stalks, which assist the insect in holding on to apparently smooth surfaces.

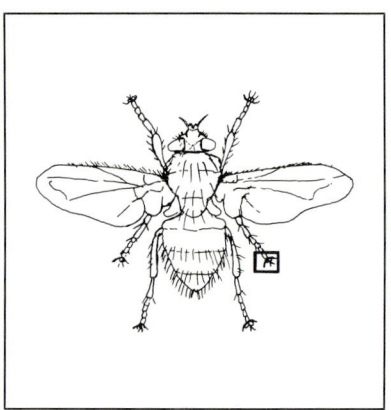

*Single seta on the thorax* [x 2960]

When the large seta is moved by contact, or by an air current, it touches small groups of setae seen in the upper left of the picture and the movement is registered by small nerves in the vicinity. This type of mechanical receptor is common in dipterous flies and many other insects.

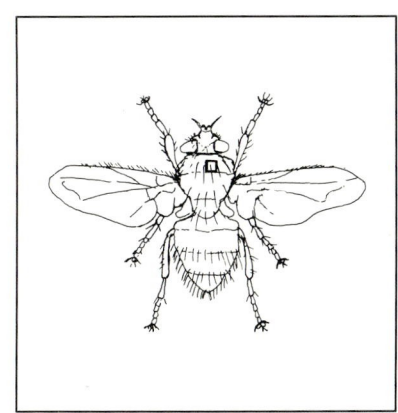

# Blowfly

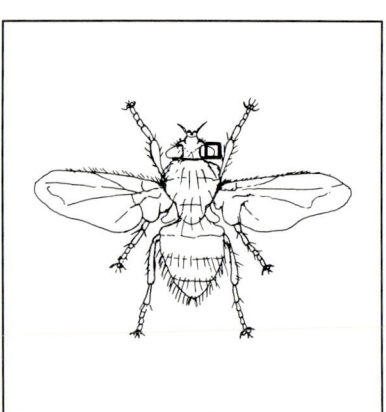

*Part of compound eye* [x 275]

The eye of a fly is composed of many small units called ommatidia. Each ommatidium has a single lens and its own nerve supply. The image the fly sees through its compound eye is probably registered in the brain as a mosaic. A mosaic image is particularly efficient for detecting movement and enables the fly to react to the slightest movement in its surroundings.

# Parasitic fly

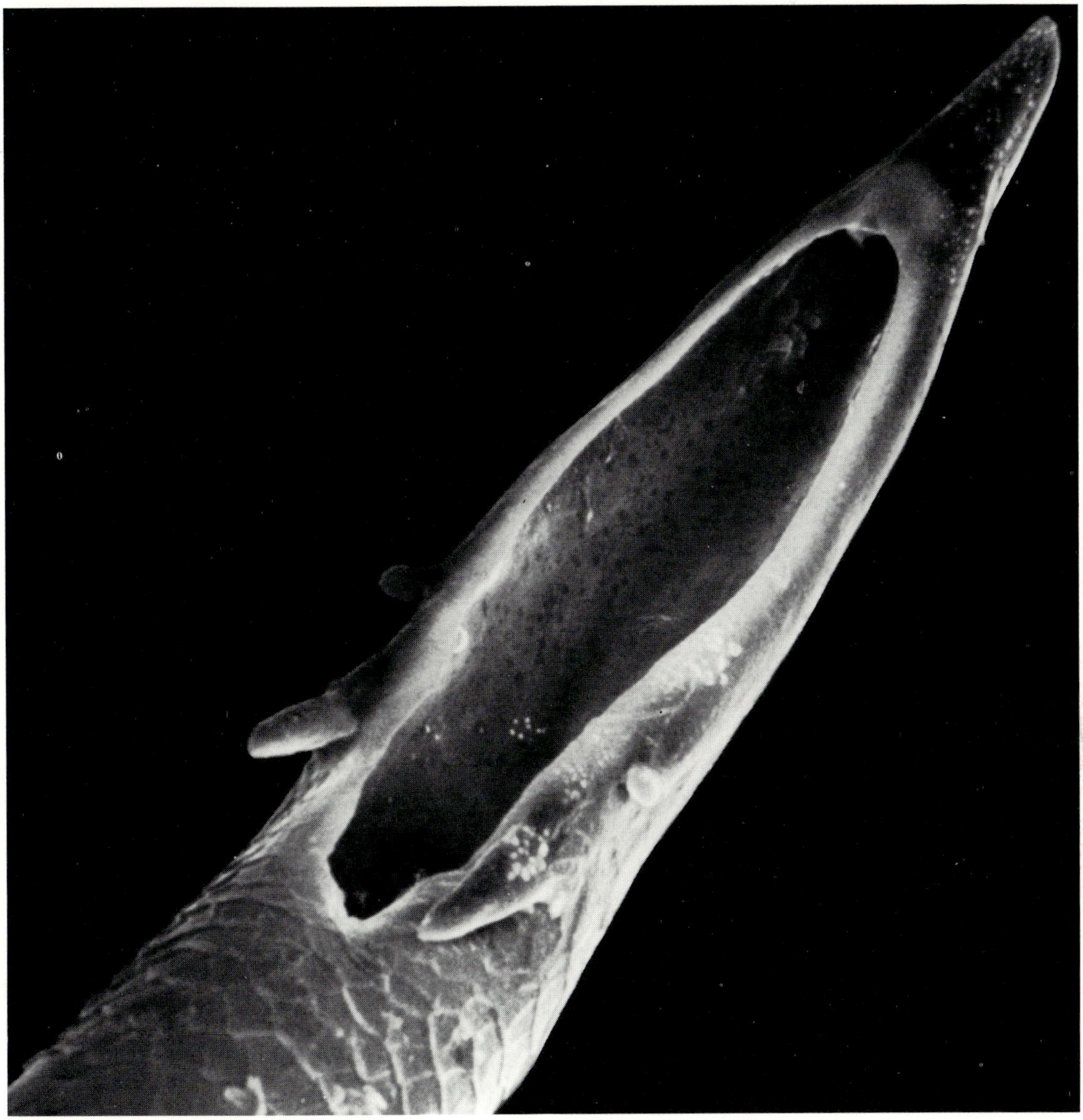

*Part of an egg* [x 980]

This is the egg of a tropical African fly, genus *Stylogaster*, of the family Conopidae. The female fly, which is about 20 mm in length, injects a single egg into the cuticle of its host, which is usually a fly or cockroach; the egg is held in place by backwardly pointing hooks. The larva emerges into the body of the host through a membranous sac, the position of which can be seen as the dark, oval-shaped area along the upper third of the egg. The larva completes its development by feeding on the living tissues of its host.

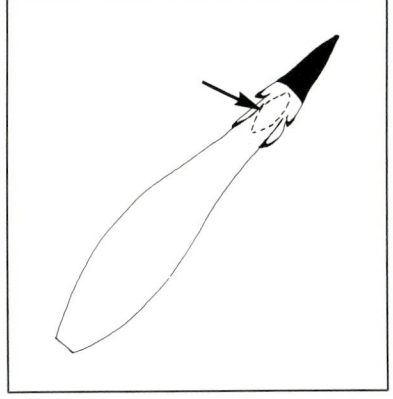

# Other invertebrates

Apart from the insects, invertebrate animals comprise an assemblage of very diverse groups. Some, like the spiders, mites and crustaceans, resemble insects in the possession of a hard exoskeleton and jointed limbs. The other groups represented here — the molluscs, echinoderms, parasitic worms and protozoans — are fundamentally different in the basic structure of their bodies, but many have hard structures that lend themselves to scanning electron microscopy. This is especially important in the case of the microscopic foraminifera whose intricate shells are abundant in both fossil and recent marine deposits.

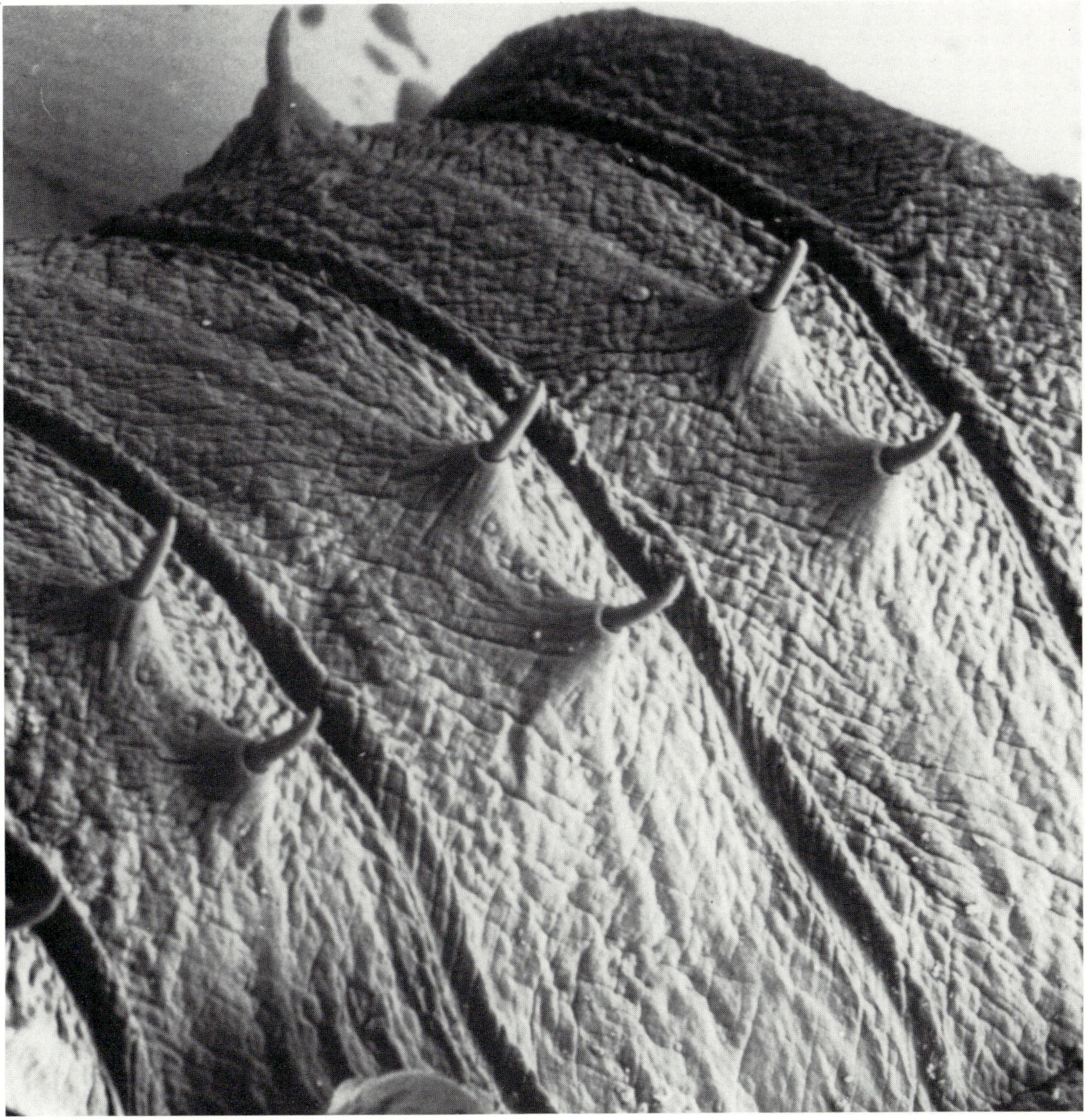

*Skin and chaetae* [x 72]

The skin of an earthworm feels rough because of these bristles or chaetae of which usually four pairs occur on each segment. The chaetae act like a ratchet in allowing forward but not backward movement as the worm extends and contracts itself. Small circular sensory organs can be seen between the chaetae and extending round the centre of each segment.

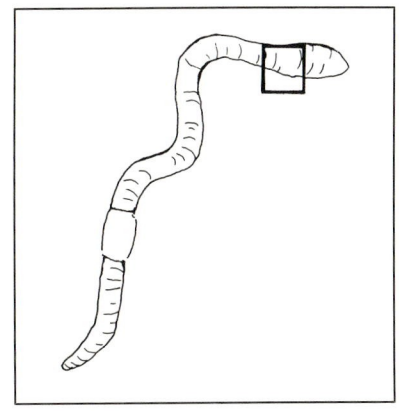

# Earthworm

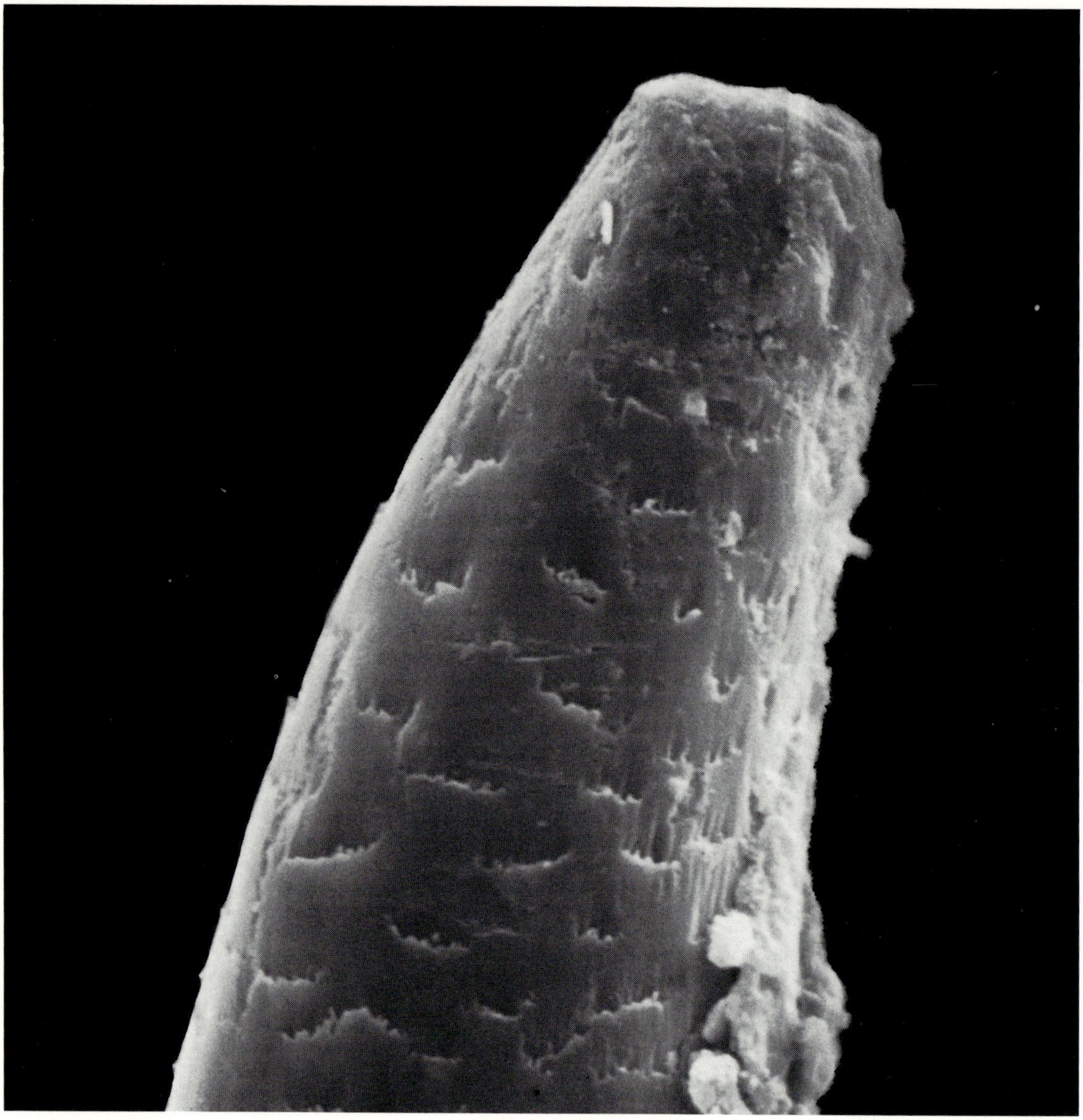

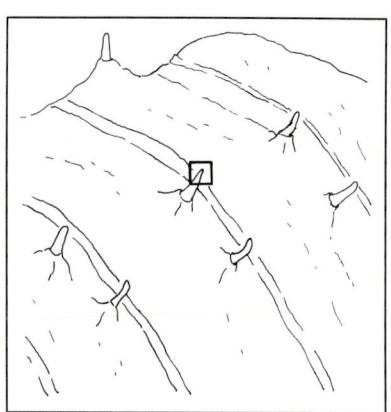

*Detail of a chaeta* [x 2500]

The chaetae of earthworms, like the more prominent ones of some other annelid worms such as the marine ragworms, consist of chitin. This is similar to, but not quite identical with that in the cuticle of arthropods such as insects, suggesting, along with the segmentation and other features, a common origin of the arthropods and the annelid worms.

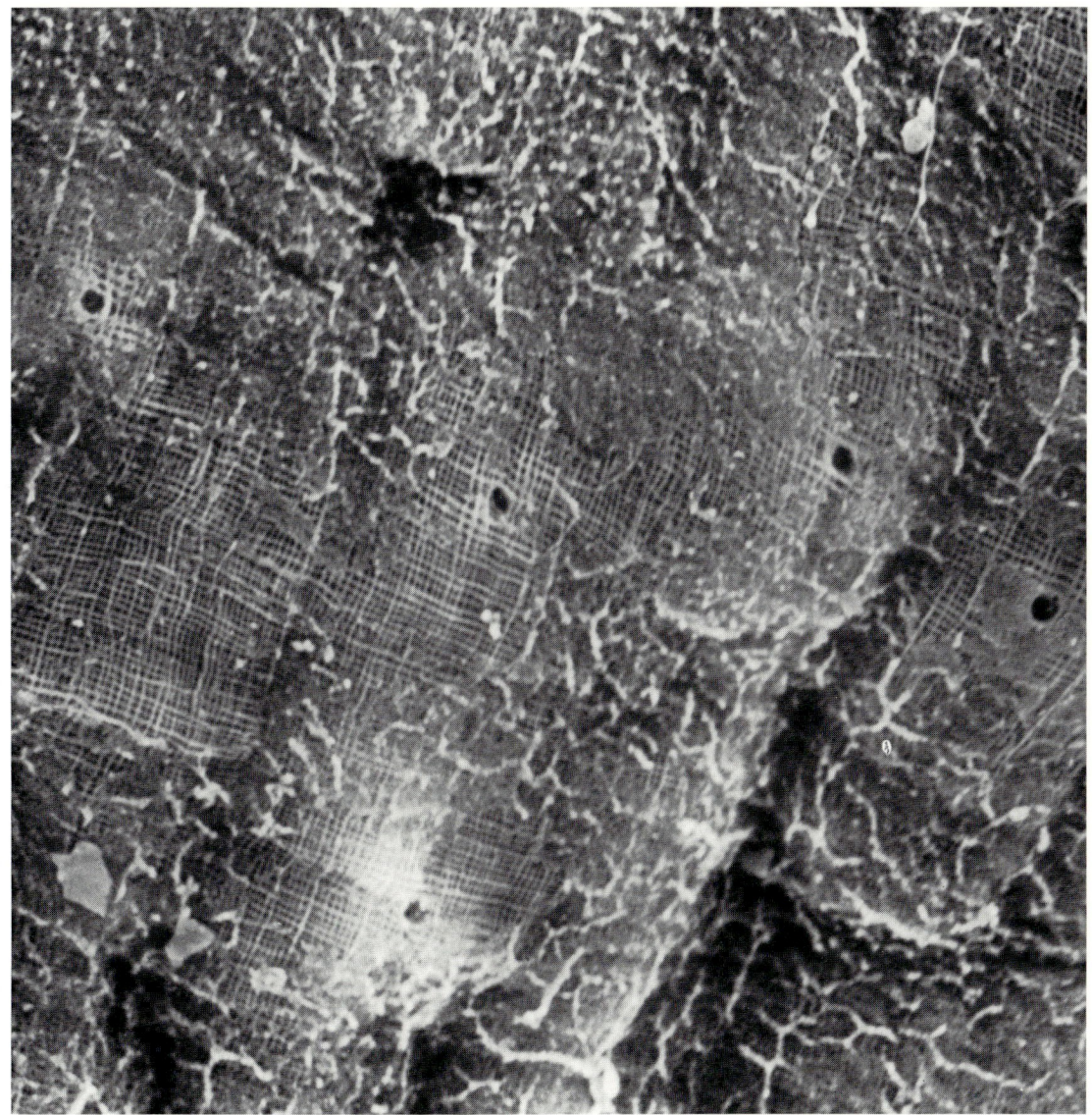

*Detail of skin of* Aporrectodea longa [x 2900]

The mode of locomotion of an earthworm requires that
the skin should be capable of extension and
contraction in both the longitudinal and radial axes.
The network of fibres seen in this picture is likely to
consist of collagen, a fibrous protein also found in the
ligaments, cartilage and bone of vertebrates.

# Nematode worm

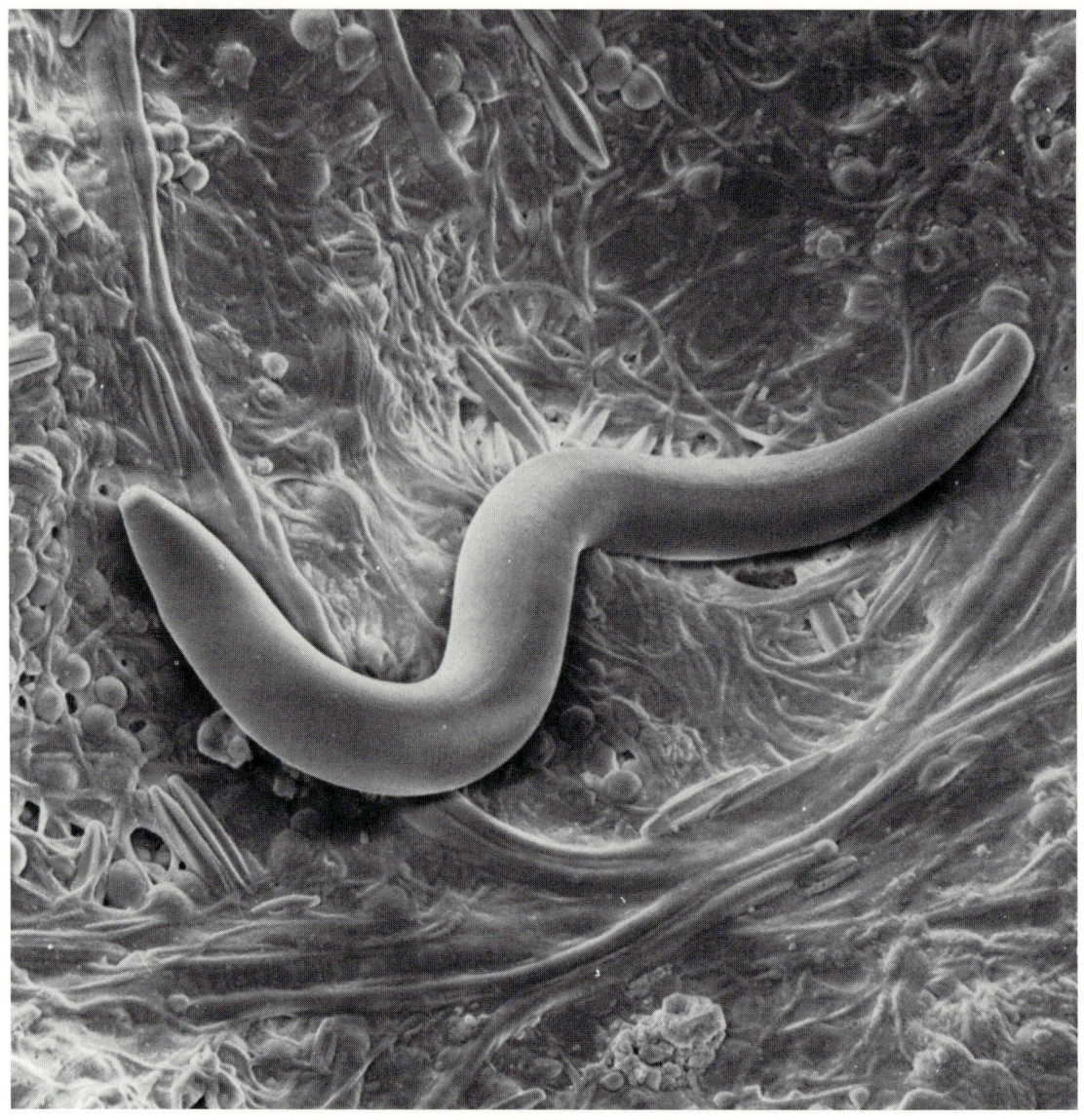

*A larval soil nematode* [x 650]

Nematode worms, also known as round worms and eel-worms, are unsegmented animals not closely related to other worms. Many are small or microscopic and occur in very great densities, and in a great diversity of species, in most aquatic or damp habitats. This one is surrounded in the soil by single-celled algae and diatoms which constitute the diet of many soil nematodes.
Photo by J.A. Sargent, Agricultural Research Council.

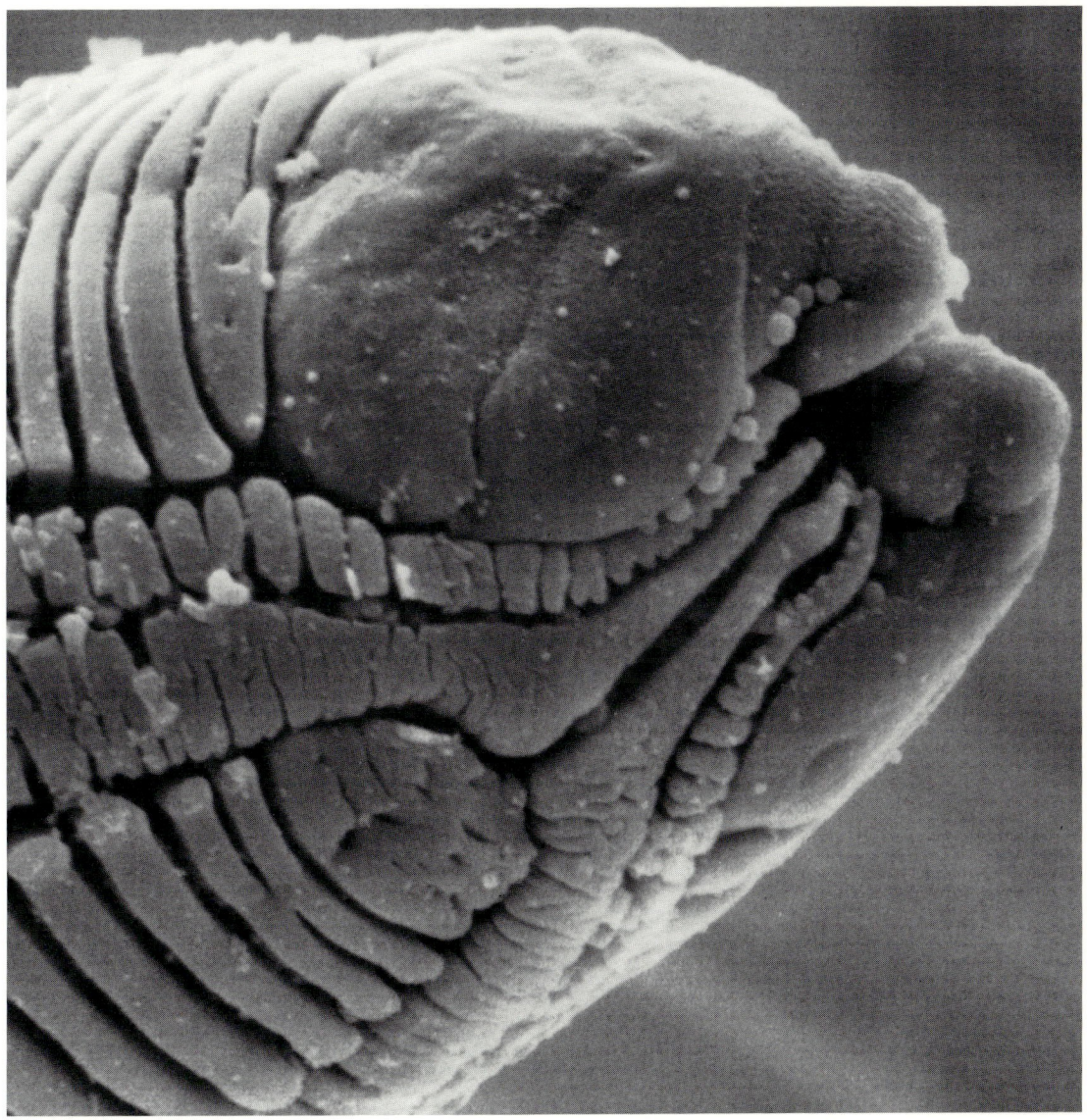

*Head of a parasitic round worm* [x 1720]

Many nematode worms are parasitic, living especially
in the gut of other animals. This one, a species of
*Acuaria* about 8 mm long, came from the gizzard of a
bird of paradise in New Guinea. Nematodes of this
group are characterized by an unusually complex
structure of the cuticle extending varying distances
behind the head.

# Nematode worm

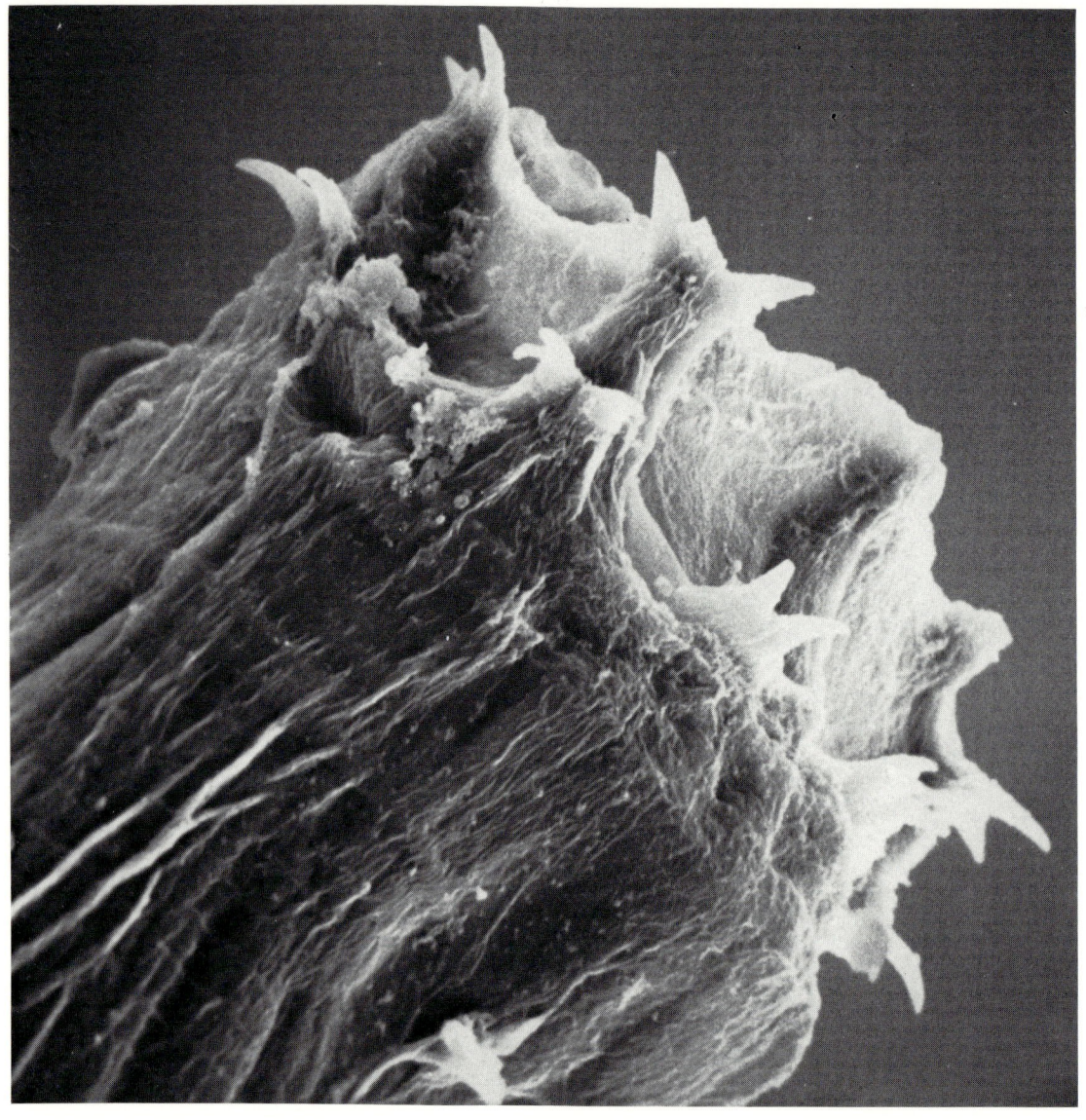

*Head of a parasitic round worm* [x 3430]

This specimen of *Grassenema procaviae*, about 2·5 mm long, was one of thousands found in the stomach and intestines of a tree hyrax in central Africa. It is unusual amongst parasitic nematodes in having rings of hooks around the mouth although many species have teeth or hooks within the mouth itself.

# Tapeworm

*Attachment organ or scolex* [x 290]

Tapeworms are internal parasites found attached to the lining of the gut in many different kinds of vertebrates. The method of attachment varies but the crown of hooks is common in tapeworms of birds and mammals. Tapeworms absorb nourishment through their skins and have no alimentary canal nor mouth. The 'head' or scolex is therefore concerned only with attachment and not feeding. This worm, *Amirthalingamia macracantha,* was recovered from the gut of a cormorant in the Sudan.

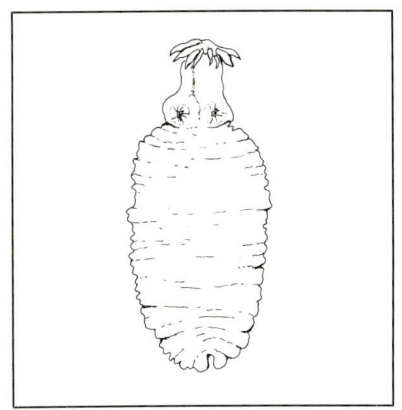

# Spider

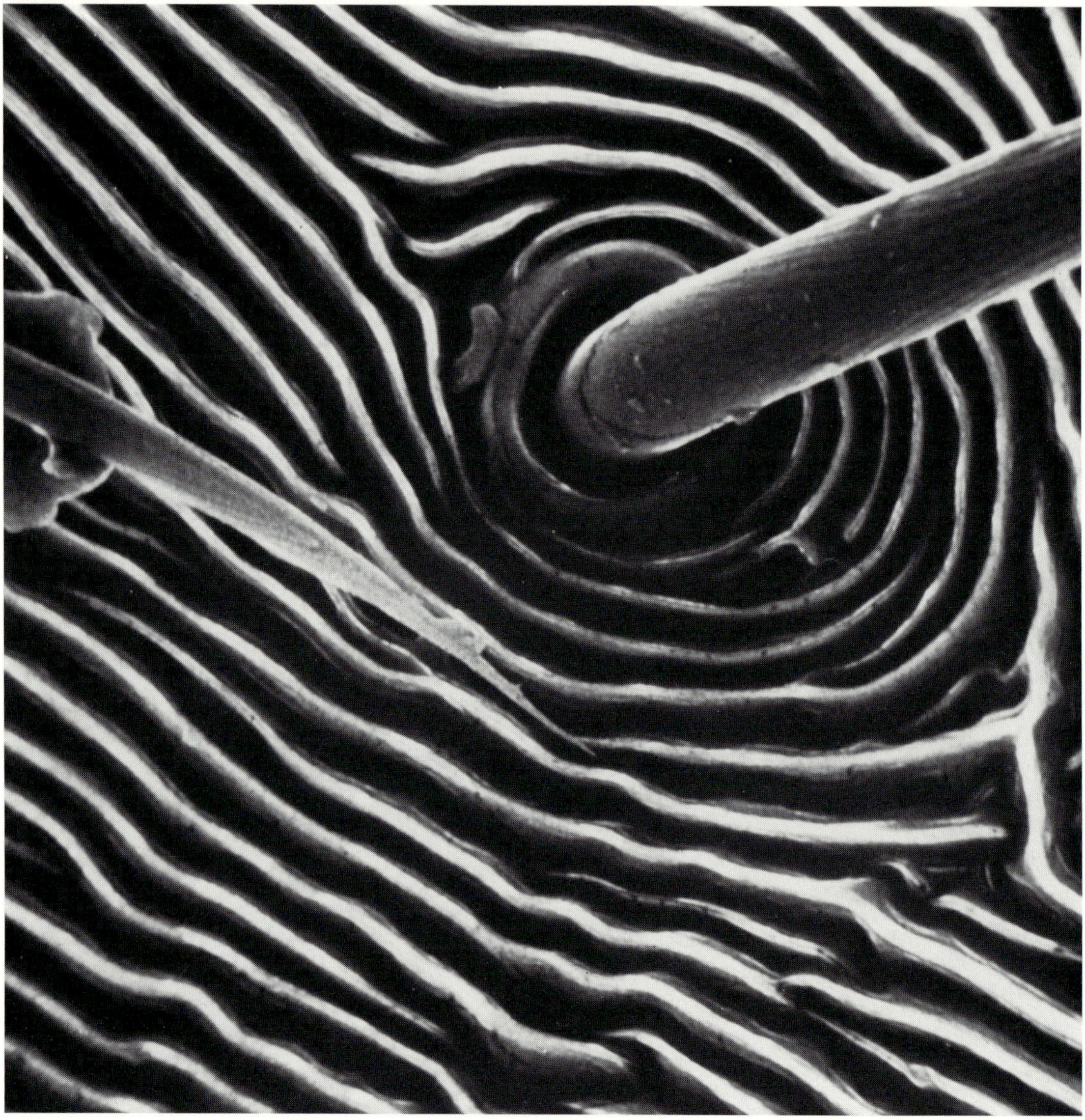

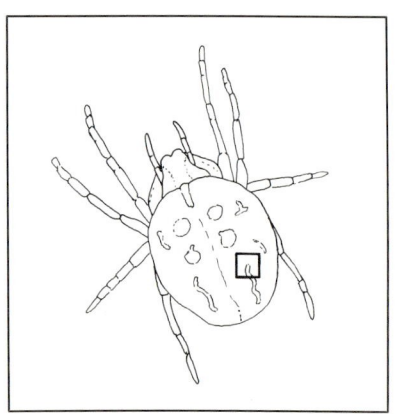

*Seta on the abdomen of a garden spider* [x 1380]

The abdomen of a spider is covered by a chitinous cuticle similar to that found in insects, but whereas in insects the cuticle forms rigid plates separated by flexible, elastic membranes, in spiders the whole cuticle of the abdomen is flexible. The pleating allows considerable expansion of the abdomen as the spider feeds or develops eggs. This species, *Araneus quadratus*, is one of the common orb-web spiders.

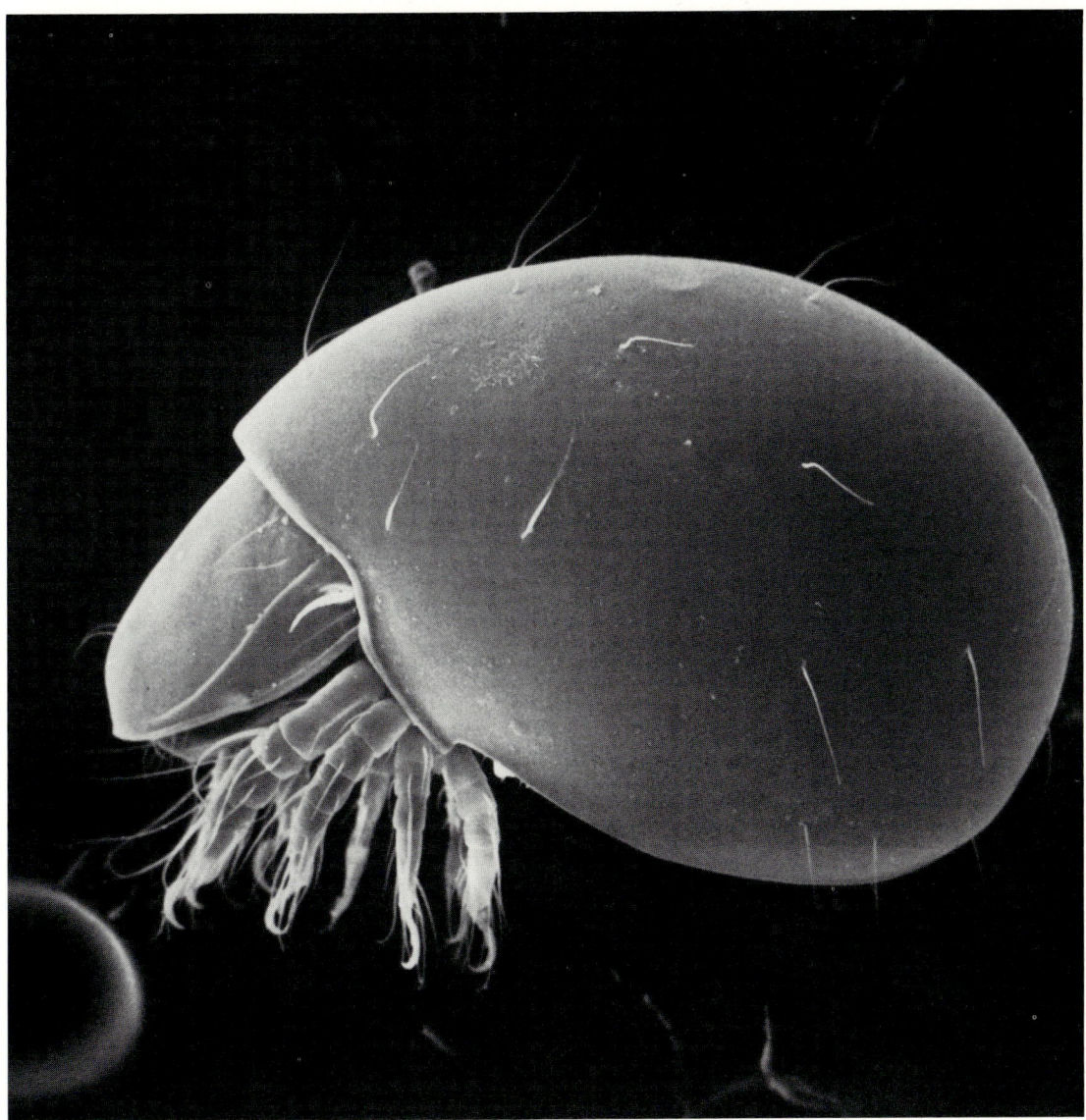

*Box mite*, Phthiracarus [x 140]

Known commonly as box mites or armadillo mites,
*Phthiracarus* species are heavily armoured as adults.
They are free-living and occur predominantly in the
upper layers of highly organic forest soils where they
play an important role in the process of organic decay
and the release of plant nutrients. In adults the legs,
mouthparts and their associated sense organs can be
withdrawn underneath the box-like body shield and
completely covered by the anterior shield.

# Mite

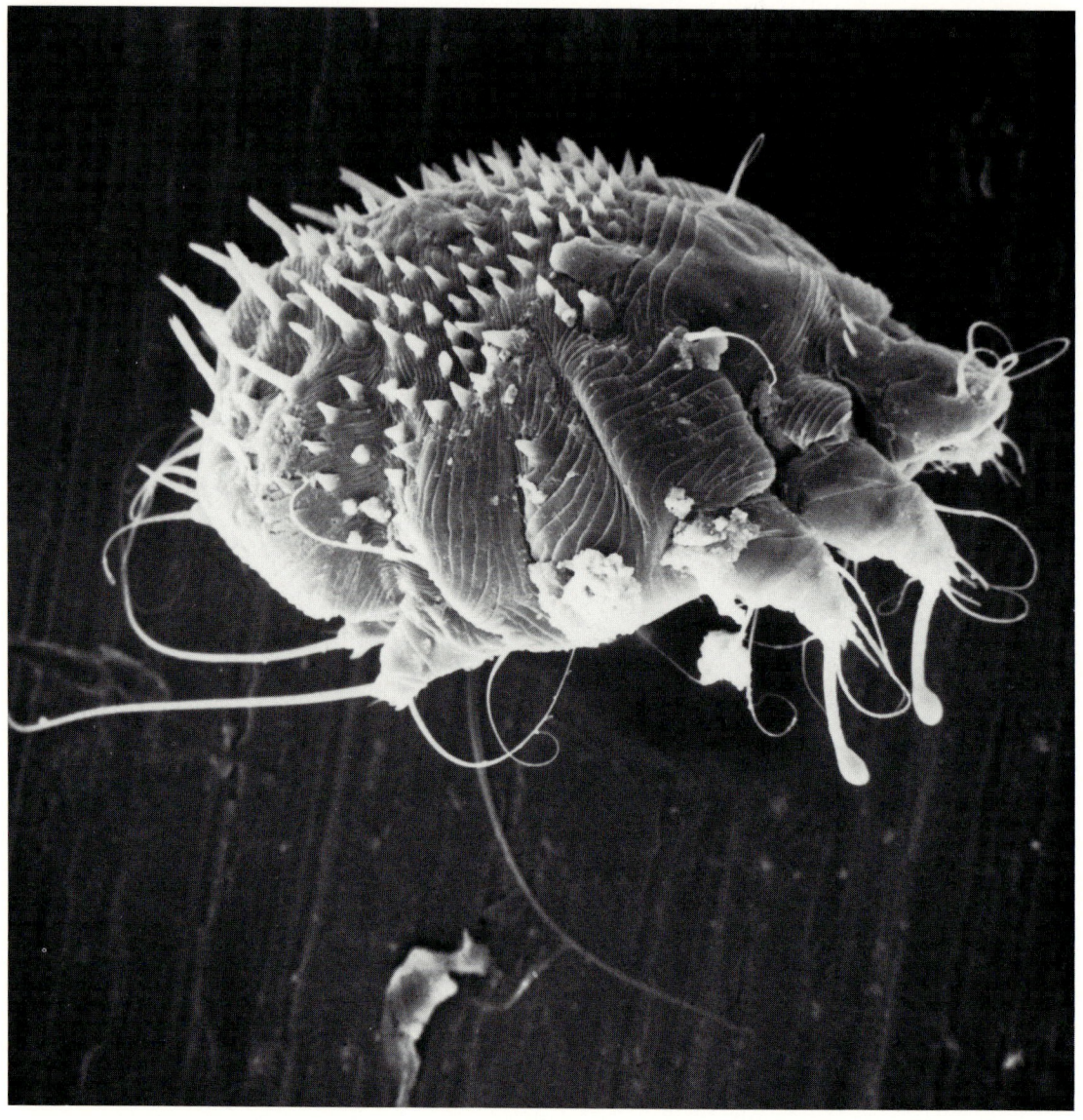

*Itch mite*, Sarcoptes scabiei [x 300]

The itch mite is a skin parasite that infests a wide
variety of mammalian hosts including man, causing
the condition known as scabies. The adult mite is
weakly armoured and the rather short legs are
provided with terminal suckers which the mite uses to
fix itself to the inside of burrows made in the surface
of the skin. Eggs are laid in the burrow and the
resulting larvae migrate to adjacent hair follicles.

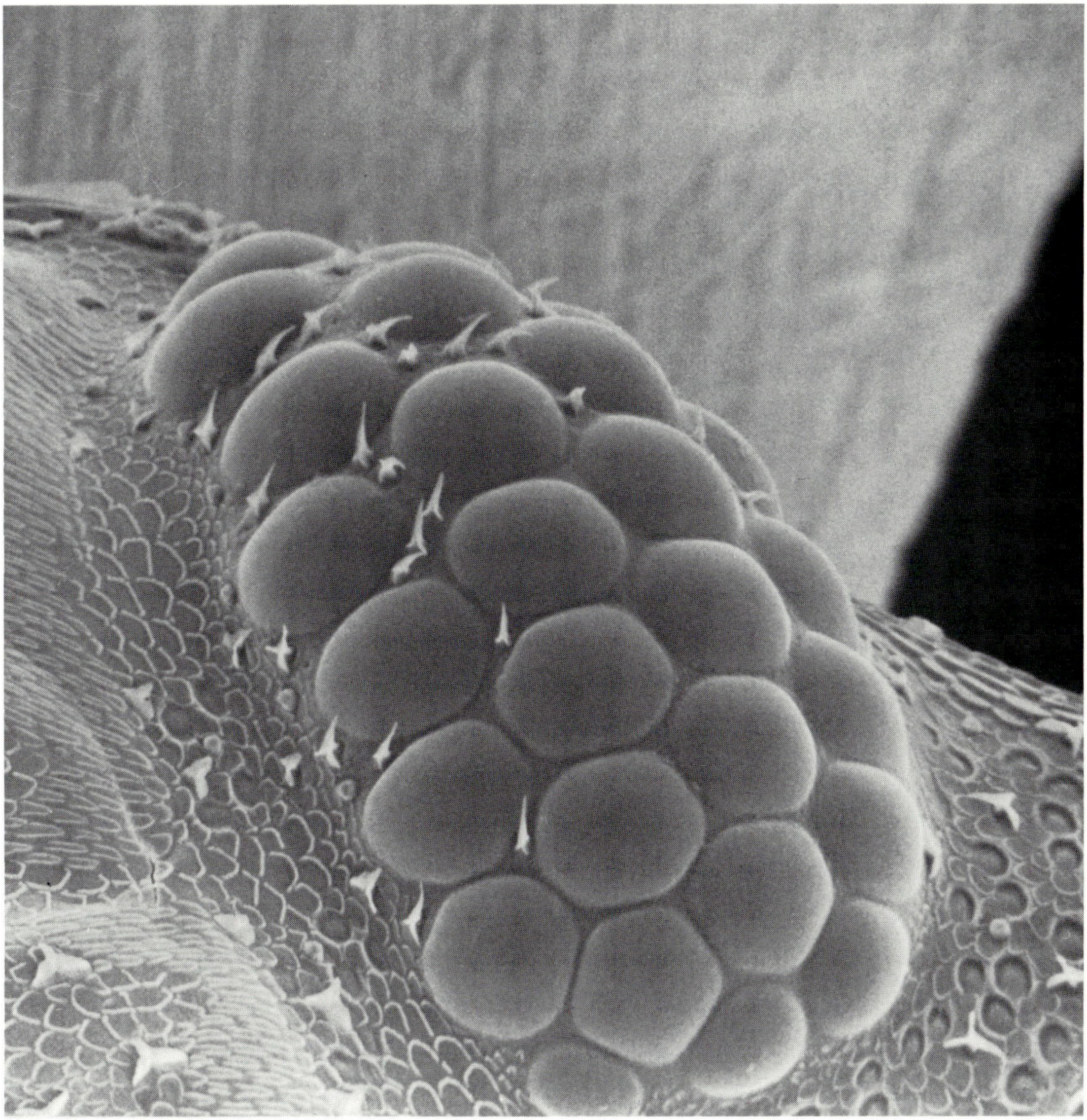

*Compound eye of* Porcellio scaber [x 260]

Woodlice, like other crustaceans such as crabs and
shrimps, share many features with insects, spiders
and mites, particularly the armoured skin and jointed
appendages. The individual elements of the compound
eye are interspersed with sensory setae similar to
those in other parts of the body.

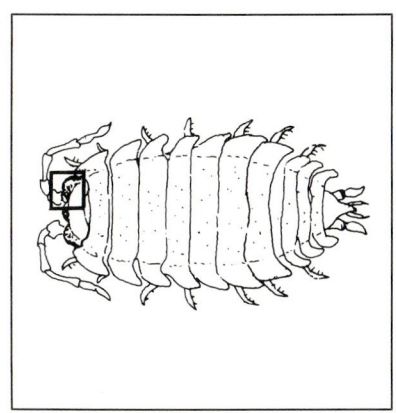

# Woodlouse

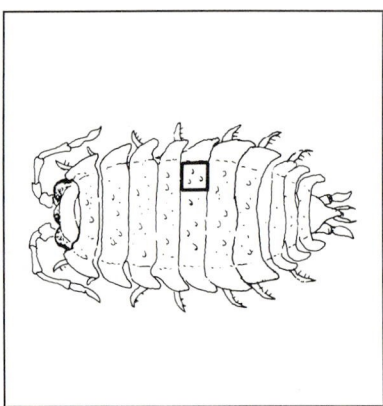

*Upper surface* [x 3160]

The scale-like plates on the back of a woodlouse are in turn covered by microscopic semicircular scales interspersed with slender sensory projections known as tricorns. These are probably touch receptors important in the 'thigmotactic' responses that favour the formation of clusters and the occupation of crevice-type habitats.

# Woodlouse

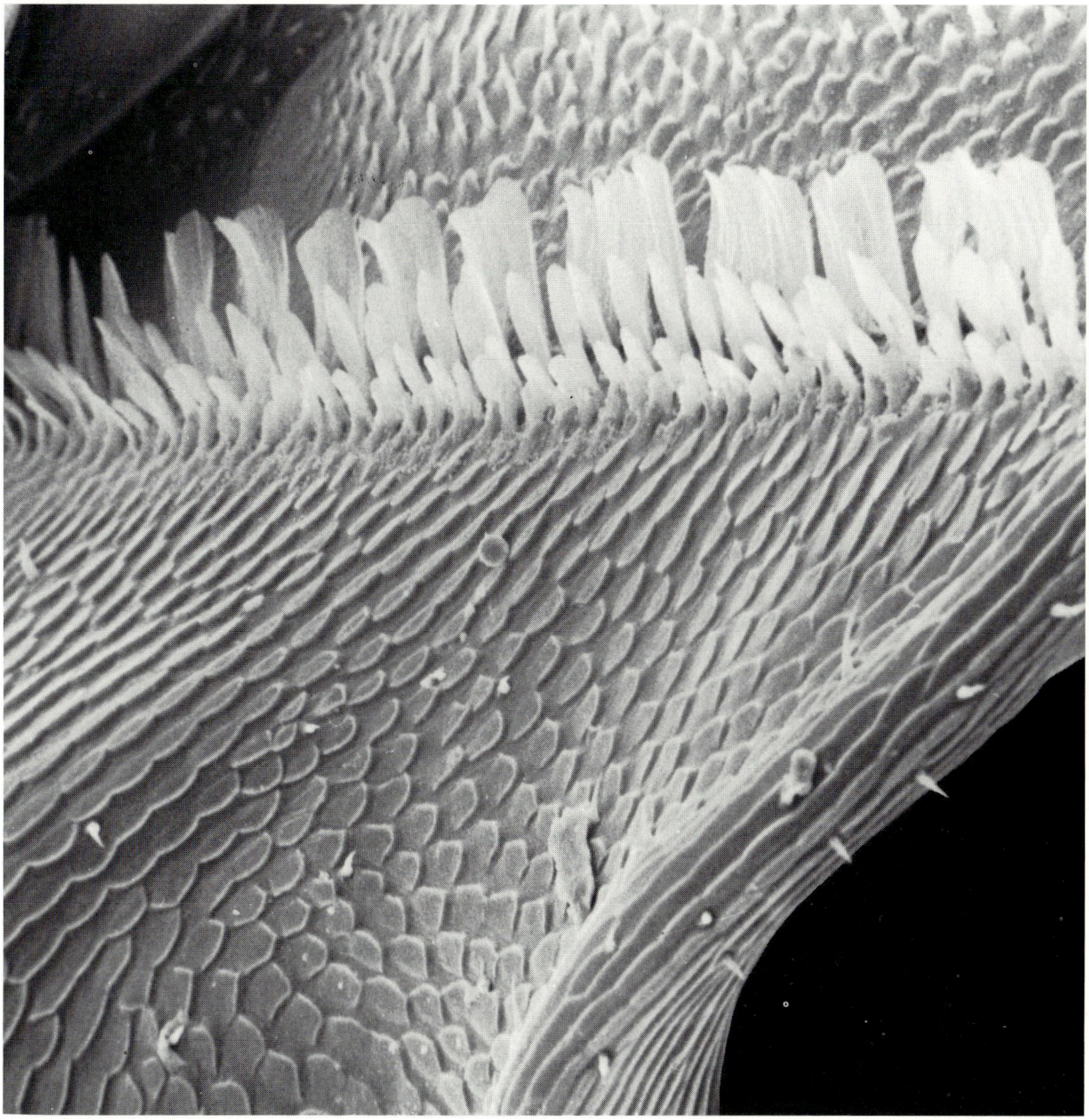

*Water channel on flank* [x 340]

Woodlice are unusual amongst crustaceans in that they live on land but they have only a limited resistance to desiccation. They have complex structures and behavioural mechanisms geared to regulating their water balance. This row of erect scales just above the leg-sockets (bottom right) forms a capillary channel that conducts water to keep the body moist when the woodlouse places its posterior appendages or 'uropods' in contact with a wet surface.

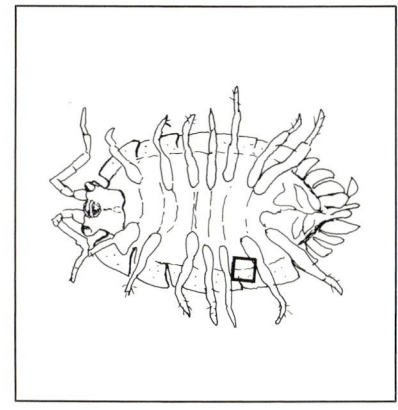

# Woodlouse

![scanning electron micrograph of woodlouse cuticle]

*Cuticle at base of leg* [x 2050]

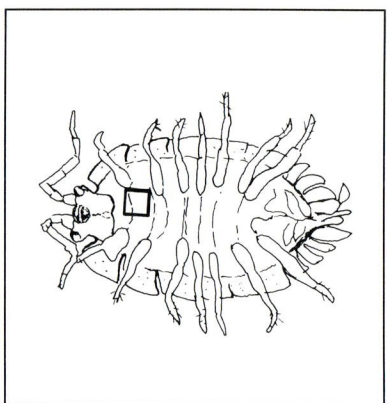

The tubercles and folding of the cuticle at the base of the legs provide an efficient compromise between strength and flexibility, allowing free articulation of the leg.

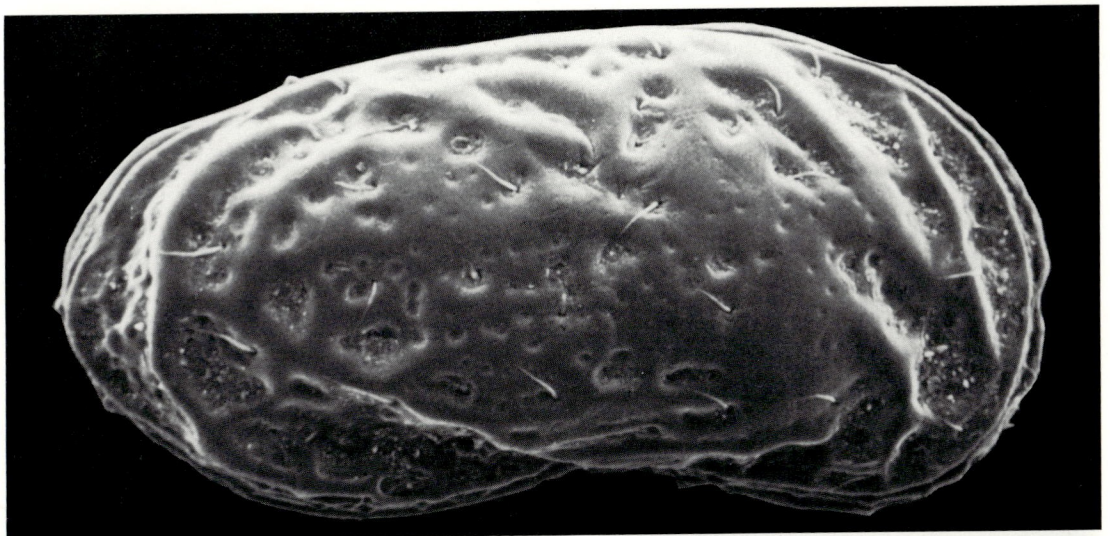

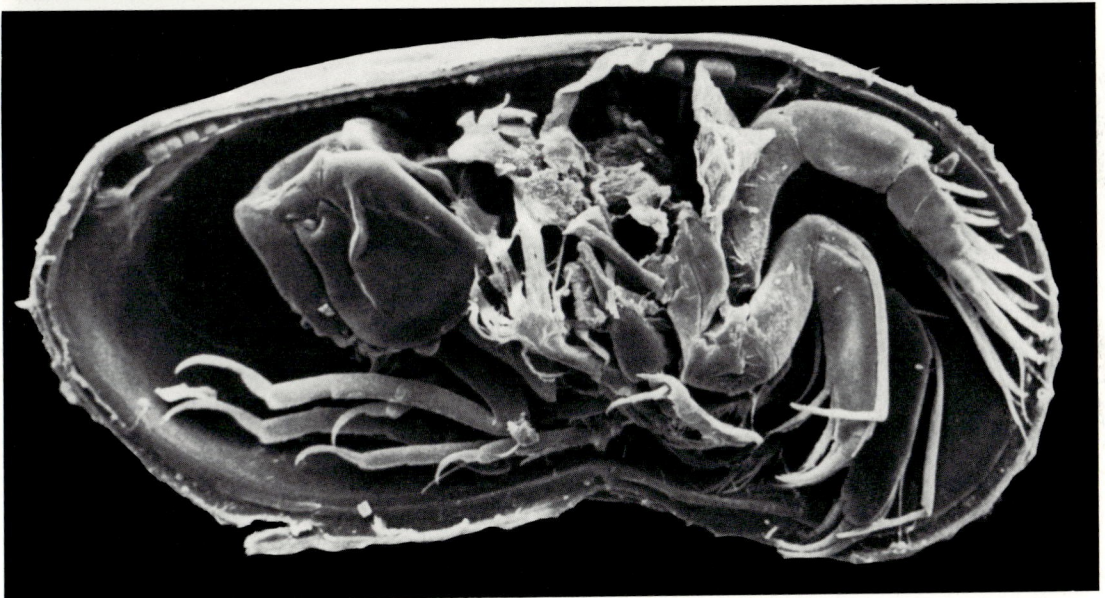

*Upper: shell closed* [x 200]
*Lower: with right valve removed* [x 200]

Ostracods are aquatic crustaceans, mostly under a
millimetre in length, whose shell is formed by a pair of
hinged valves like those of a bivalve mollusc, for
example the common mussel. The shell is of calcium
carbonate but differs from that of a mollusc in showing
no growth-lines (it is moulted and regrown as in other
arthropods), and in bearing setae. Inside, the animal is
more obviously crustacean, with complex jointed
appendages. This specimen, a male of *Callistocythere
badia*, came from seaweed on the coast of Cyprus.

# Ostracod crustaceans

a  b  c

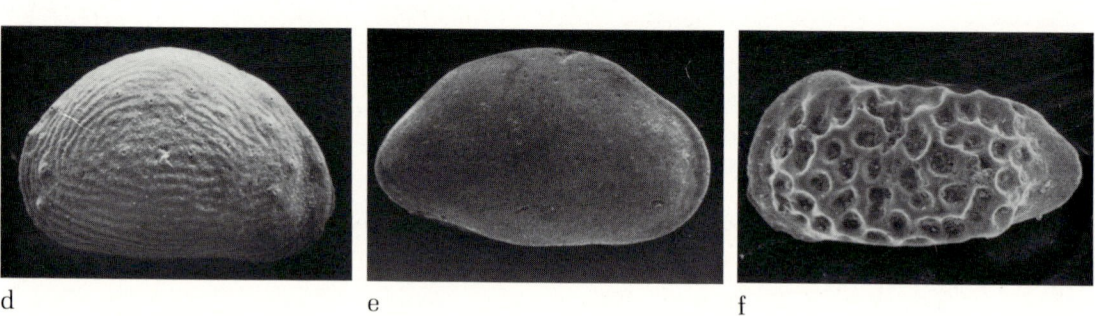

d  e  f

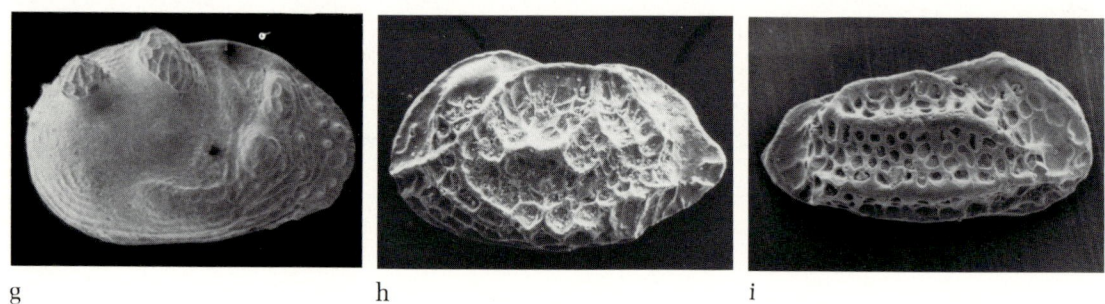

g  h  i

*A variety of fossil shells from the Jurassic Period*
*[x 40 — 100]*

Ostracods live in aquatic habitats from small
freshwater ponds to the ocean floor. The shells fossilize
readily. They first appear in sedimentary rocks of the
Cambrian period, 550 million years ago, and are very
abundant at the present day. Those shown here, from
Middle Jurassic rocks of southern England about 165
million years old, show the variety of pattern and the
detail revealed in fossil ostracods that makes them of
particular importance in dating rocks and assessing
the conditions under which they were deposited:
a, d and g were freshwater species, the rest marine.

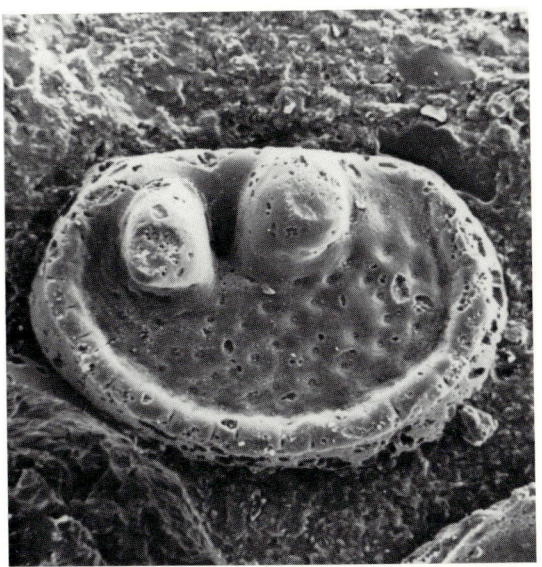

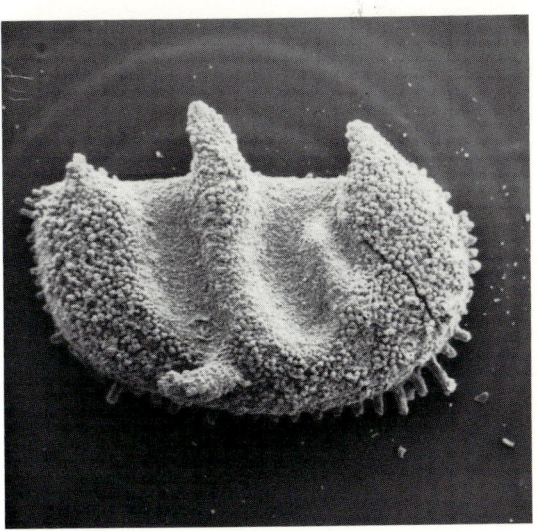

*Four fossil species from the Ordovician Period*
*[x 20 — 60]*

These shells are about 450 million years old, from Ordovician rocks of Britain, and are some of the earliest, best preserved specimens known. Some fossils consist of the impression made by the animal in fine sediment rather than the fossilized animal itself. The upper photographs are of casts made by pouring a plastic material into such natural fossil moulds. In the lower examples the calcium carbonate of the shell has been naturally replaced by silica in the course of fossilization.

# Ostracod crustaceans

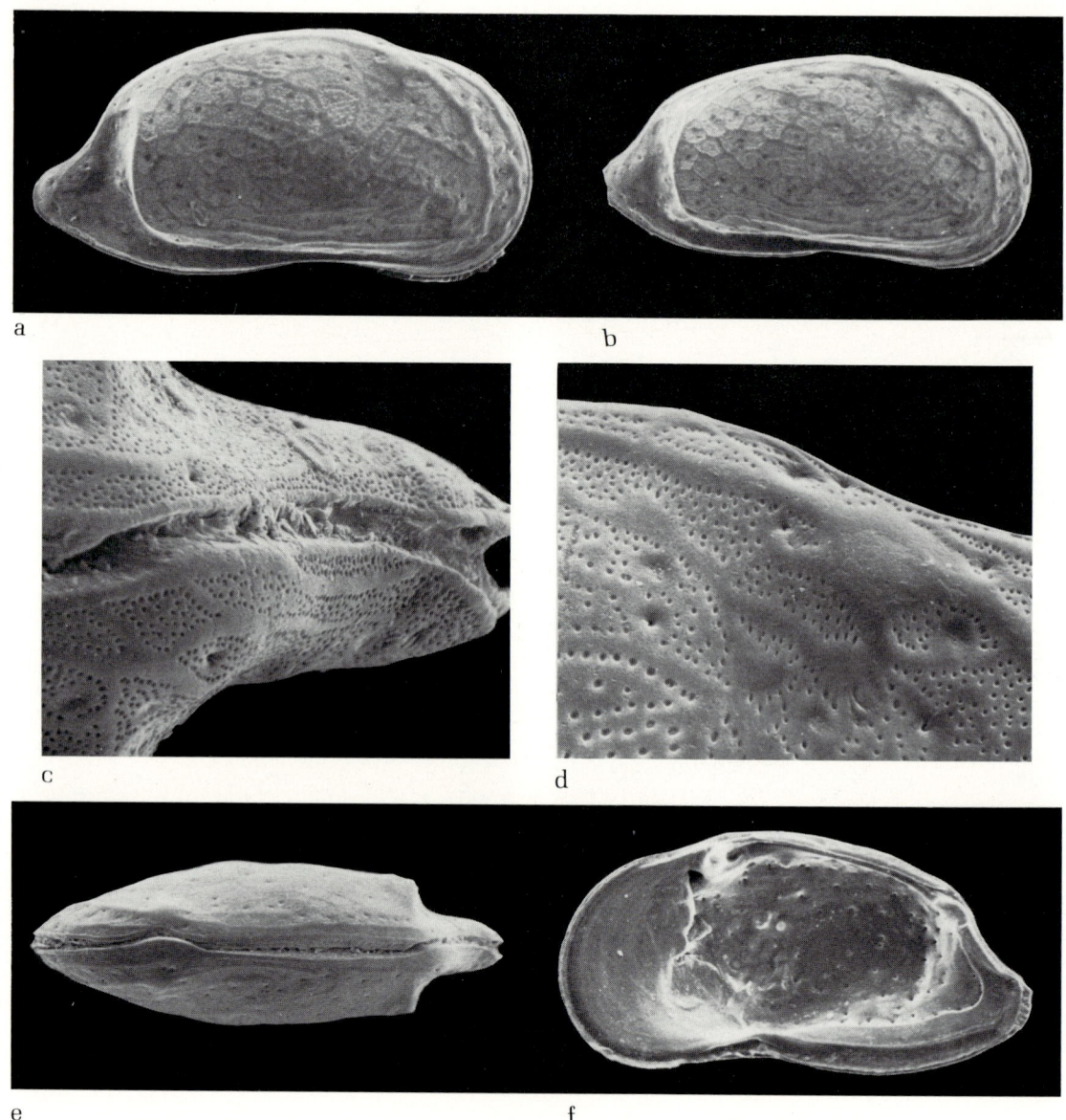

a  b  c  d  e  f

*Details of the shell of* Palaciosa cracenta

This is a living species, collected from seaweed along the coast of Ecuador. (a) Right valve, female [x 90], (b) Right valve, male [x 90]. (c) Detail of rear of shell from above [x 425]. (d) Detail of eye node — the unpatterned lump is a transparent window in the shell that overlies the animal's eye [x 475]. (e) View from above showing bivalved shell [x 95], (f) Inside of right valve showing hinge (top) [x 95].

# Garden snail

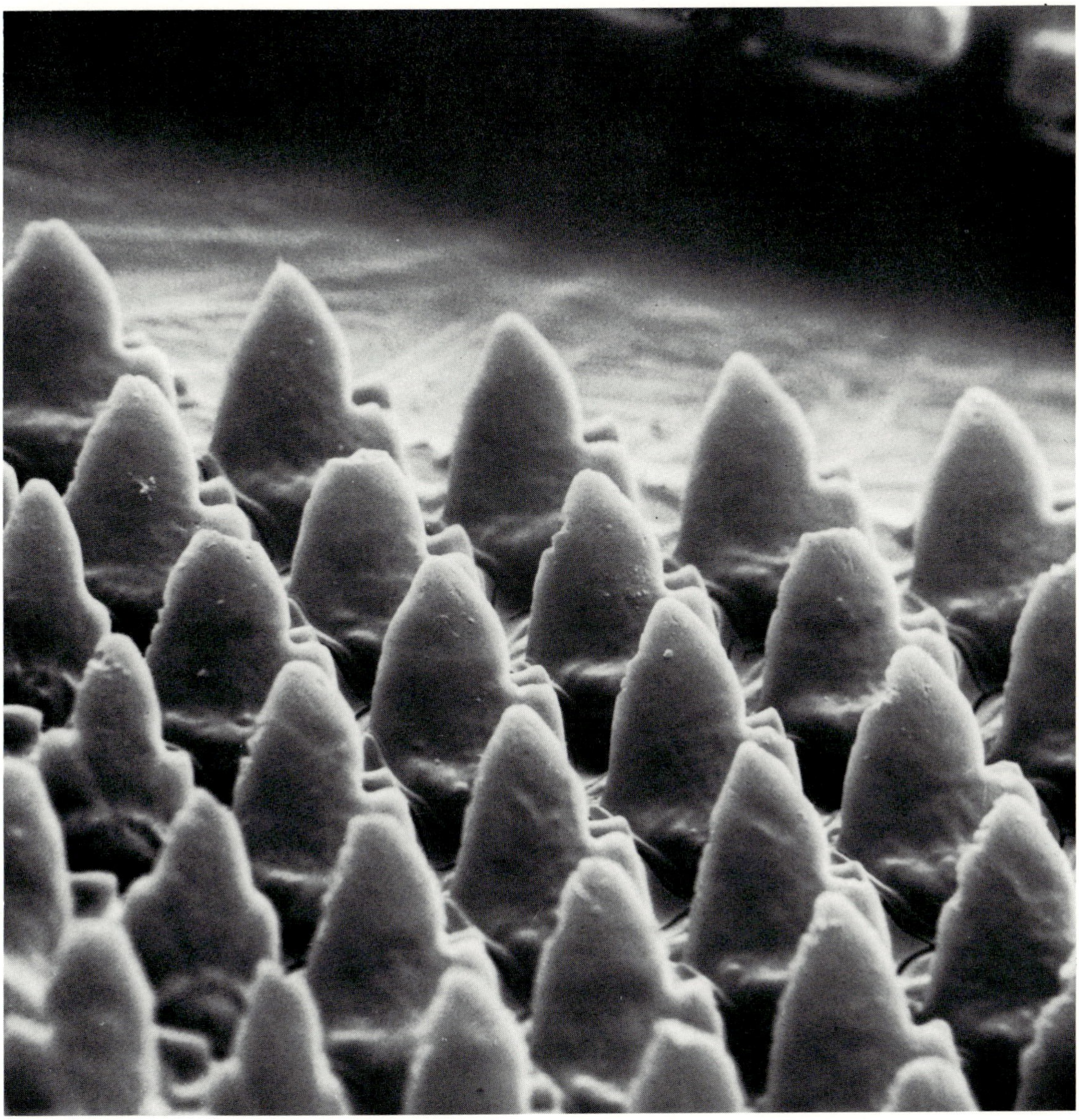

*Teeth* [x 690]

The common garden snail, *Helix aspersa*, is well known for its destruction of young seedlings and other tender plants. All snails found in the garden have teeth, which are arranged in rows on a tongue-like structure; in *Helix aspersa* there are about seventy-eight teeth in each row and about a hundred rows. The apparatus is called a radula and is used to rasp tissue from plants.

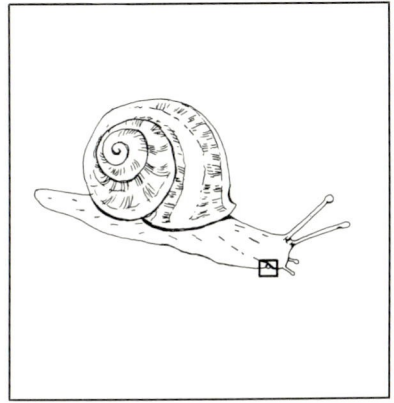

# A marine snail

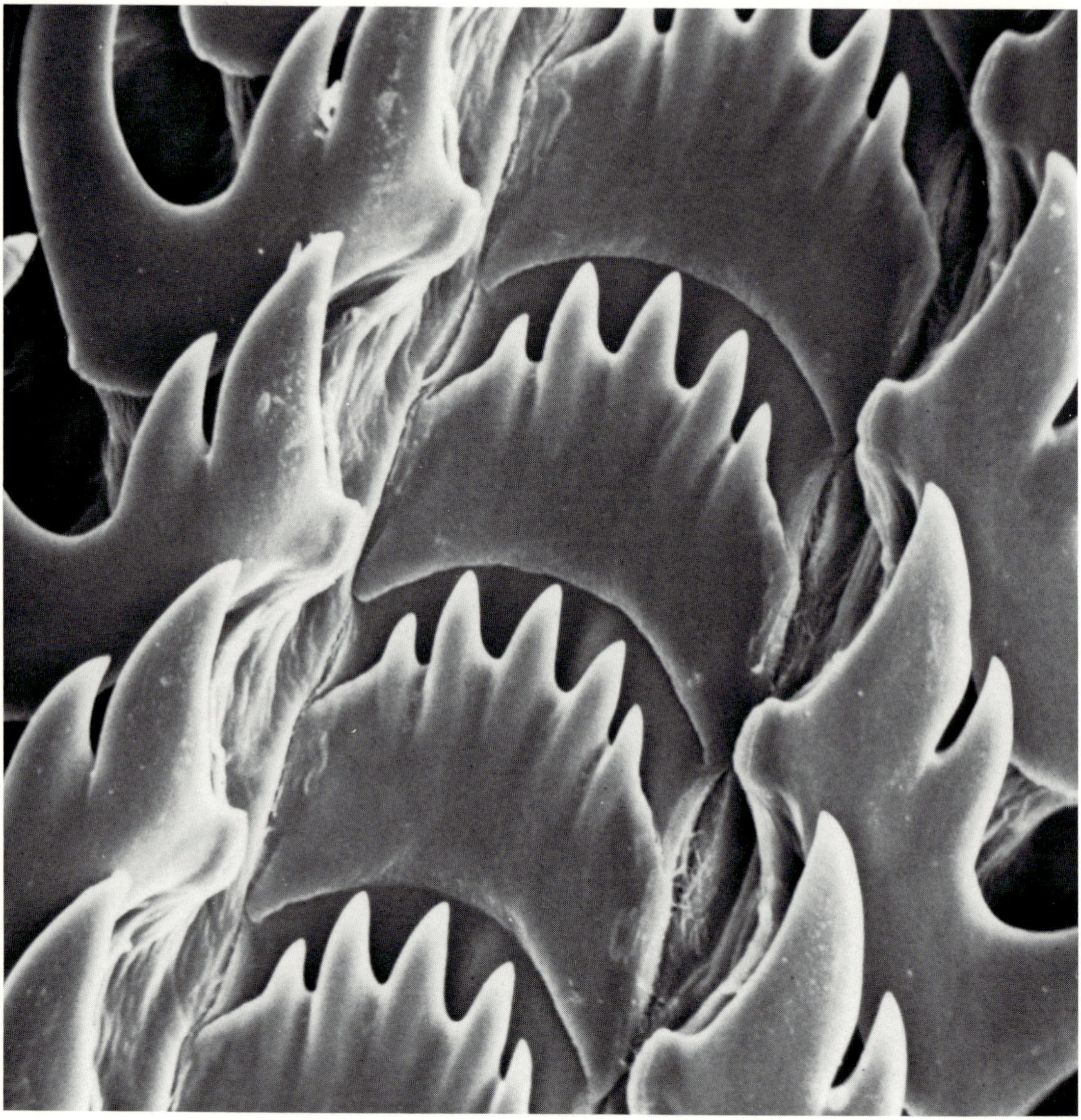

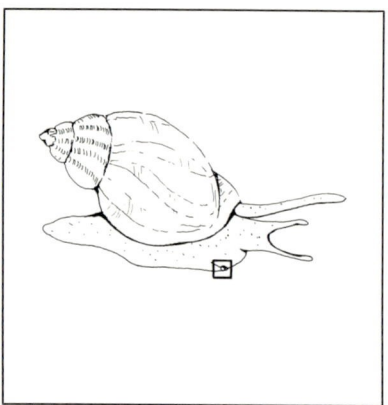

*Teeth* [x 655]

This is part of the radula of a species of small whelk (*Pisania*) that feeds on small invertebrates. As in the garden snail the radula has many teeth, but only three to a row. The whole of the radula is not used at one time, only the first few rows, the remaining part being coiled rather like a clock spring in a structure known as the radula sac. As the teeth at the front end wear and drop off the whole tongue unwinds and moves forward.

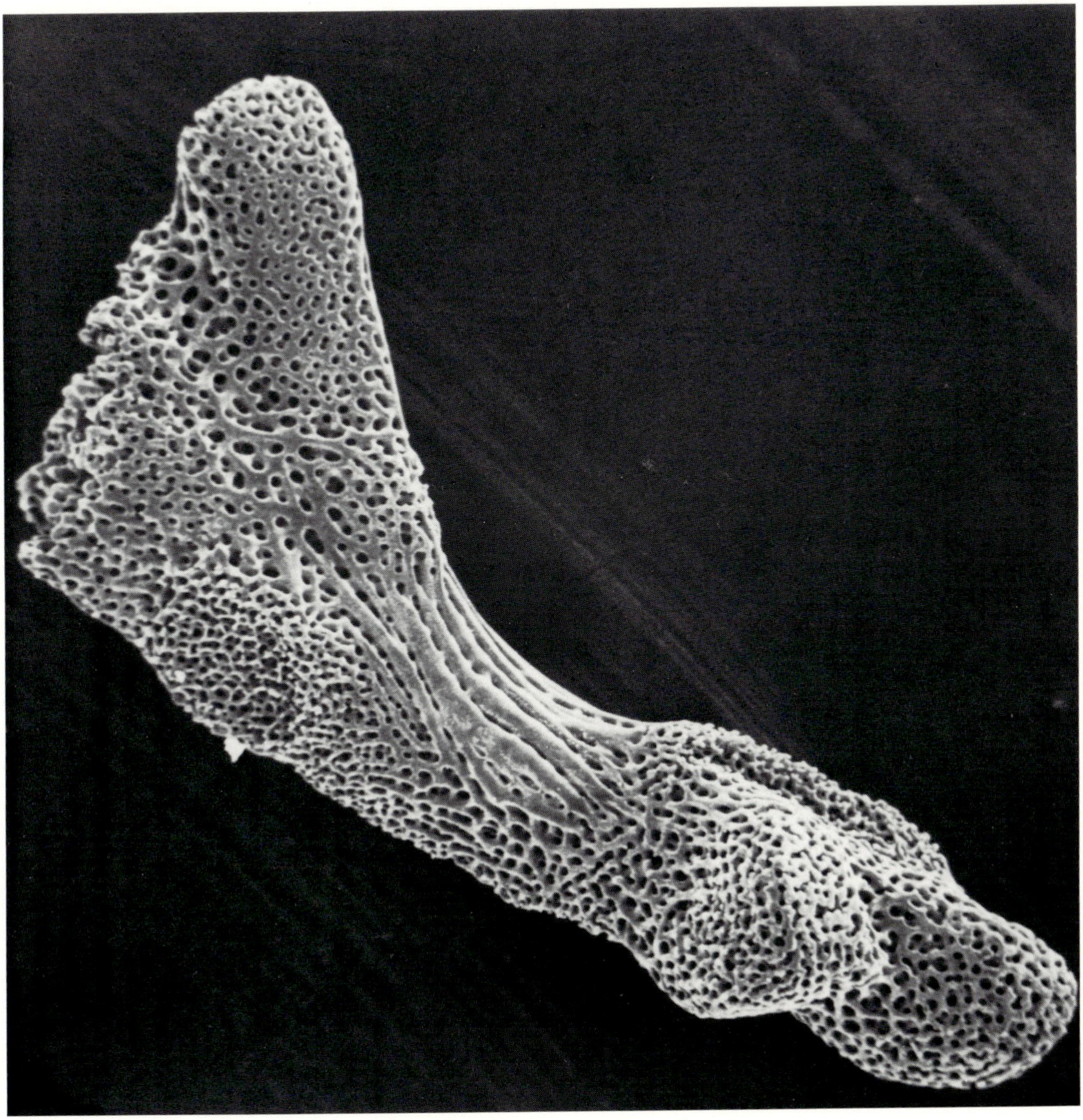

![Ossicle electron micrograph]

*Ossicle* [x 125]

Starfishes have an internal but superficial skeleton made up of a large number of small, variously shaped ossicles composed of calcium carbonate. These provide attachment for muscular and other tissues on the inside and for any surface armament. This ossicle is from a species of *Asterina* found on British shores.

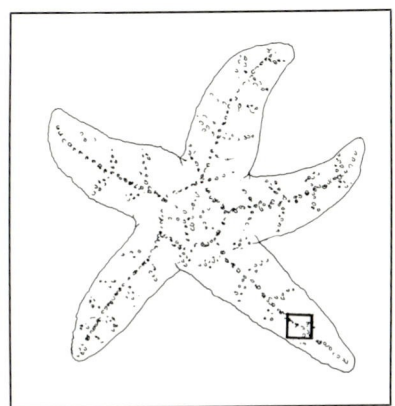

# Starfish

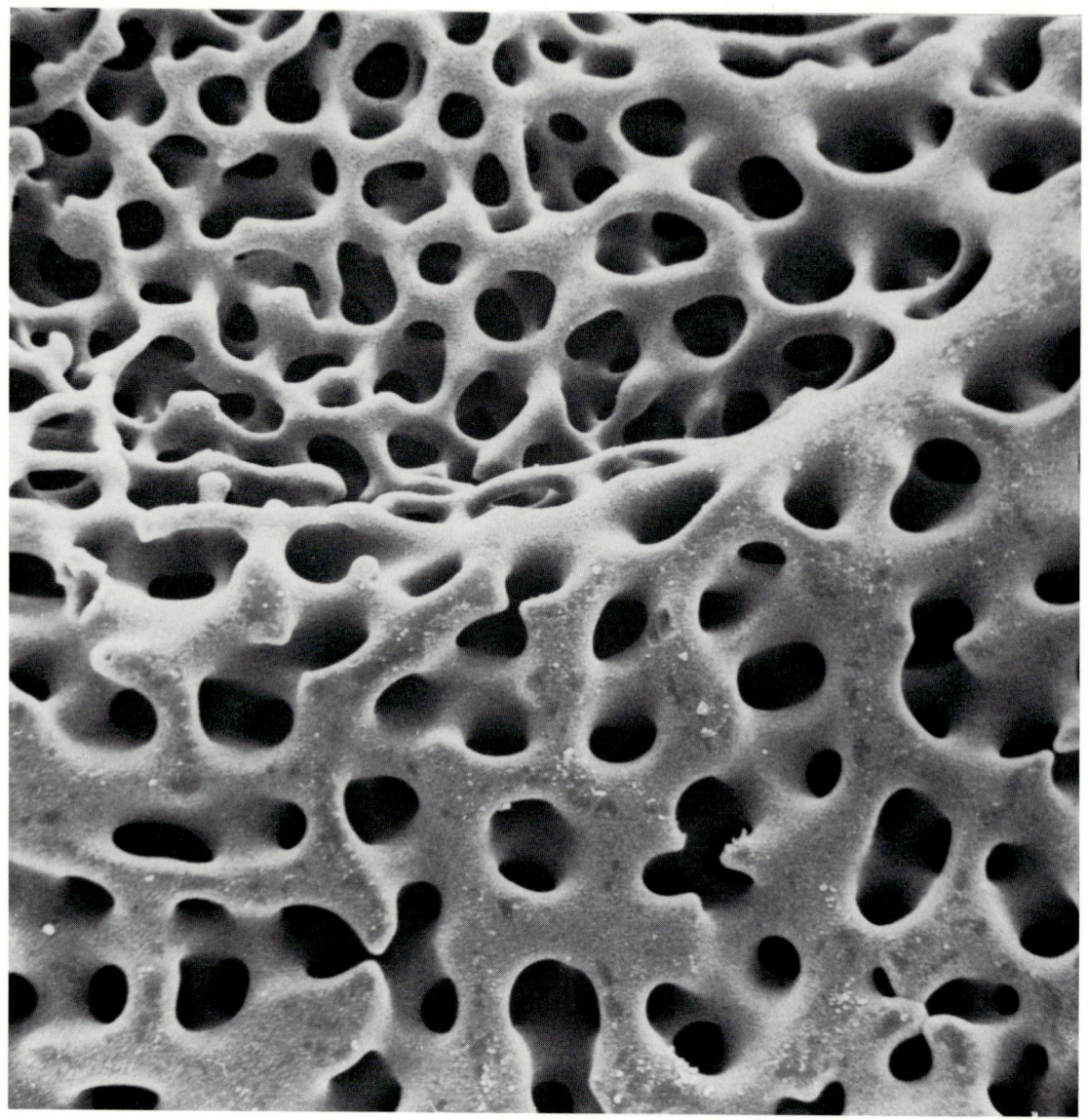

*Detail of an ossicle* [x 660]

The ossicles forming the skeleton of a starfish have a characteristic perforated structure quite different from that found in the bones of vertebrate animals.

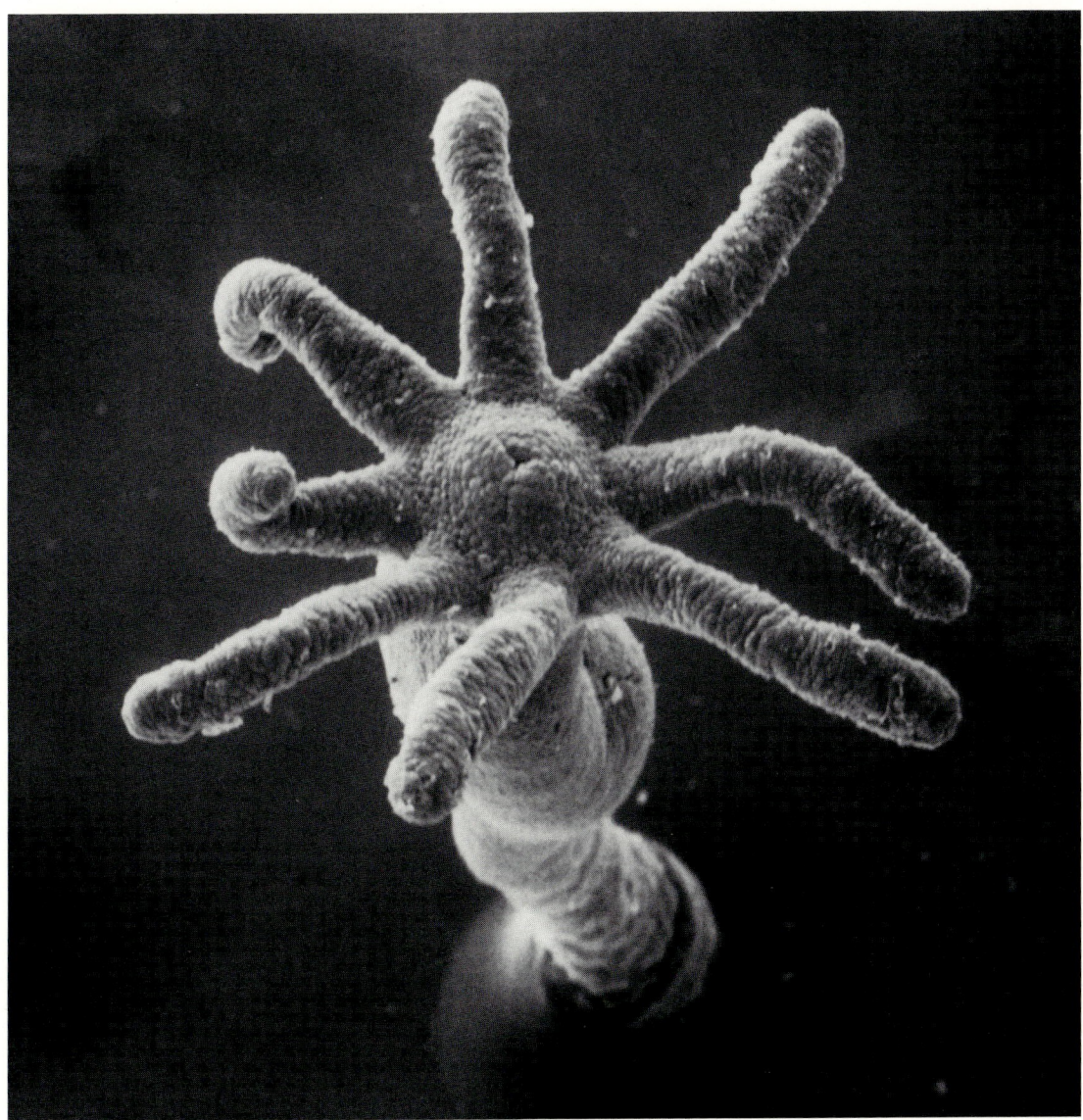

*Tentacles and mouth* [x 60]

Hydras are coelenterates, i.e. they are related to the sea anemones, corals and jellyfish, and are common in fresh water. They usually remain attached to a plant or stone and capture prey such as small crustaceans with their tentacles. The central mouth is the only opening of the digestive cavity. This animal, a specimen of *Hydra fusca*, is partially contracted. It was narcotized before fixing, and dried by the critical point method. The bulge on the column is a developing gonad.

# Hydra

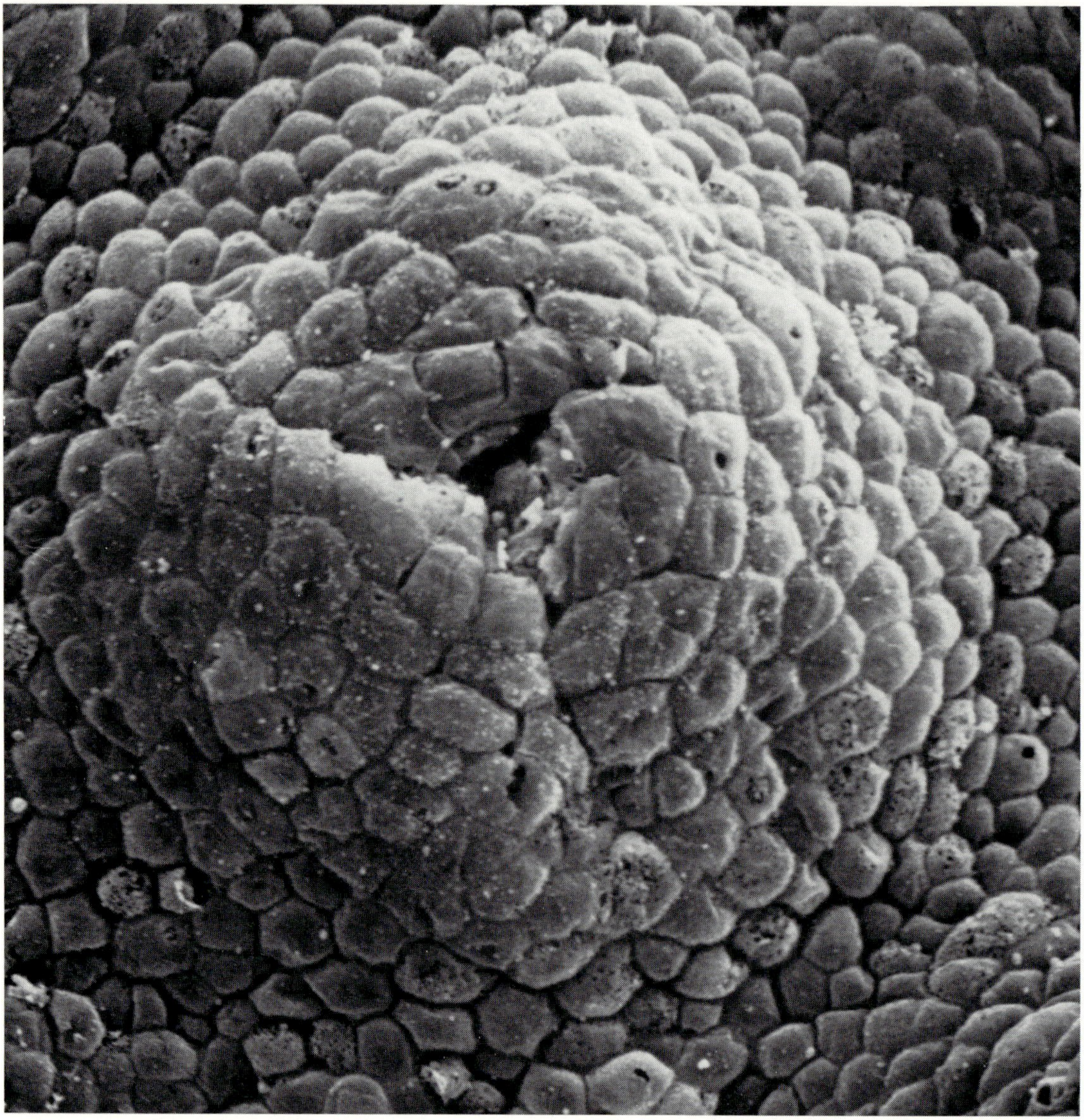

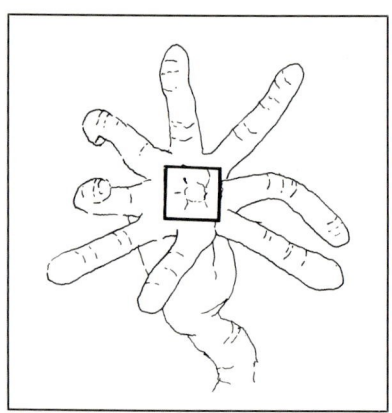

*Detail of mouth* [x 250]

The minute mouth is situated on top of a mound or hypostome. The specimen is partly contracted. In life the ectodermal cells, here seen as a knobbly crazy-paving, would be wider and flatter. The mouth distends greatly, up to about the width of the picture, when engulfing prey. Digestion takes place in the gut or enteron and the tightly-closing mouth prevents loss of the digesting meal.

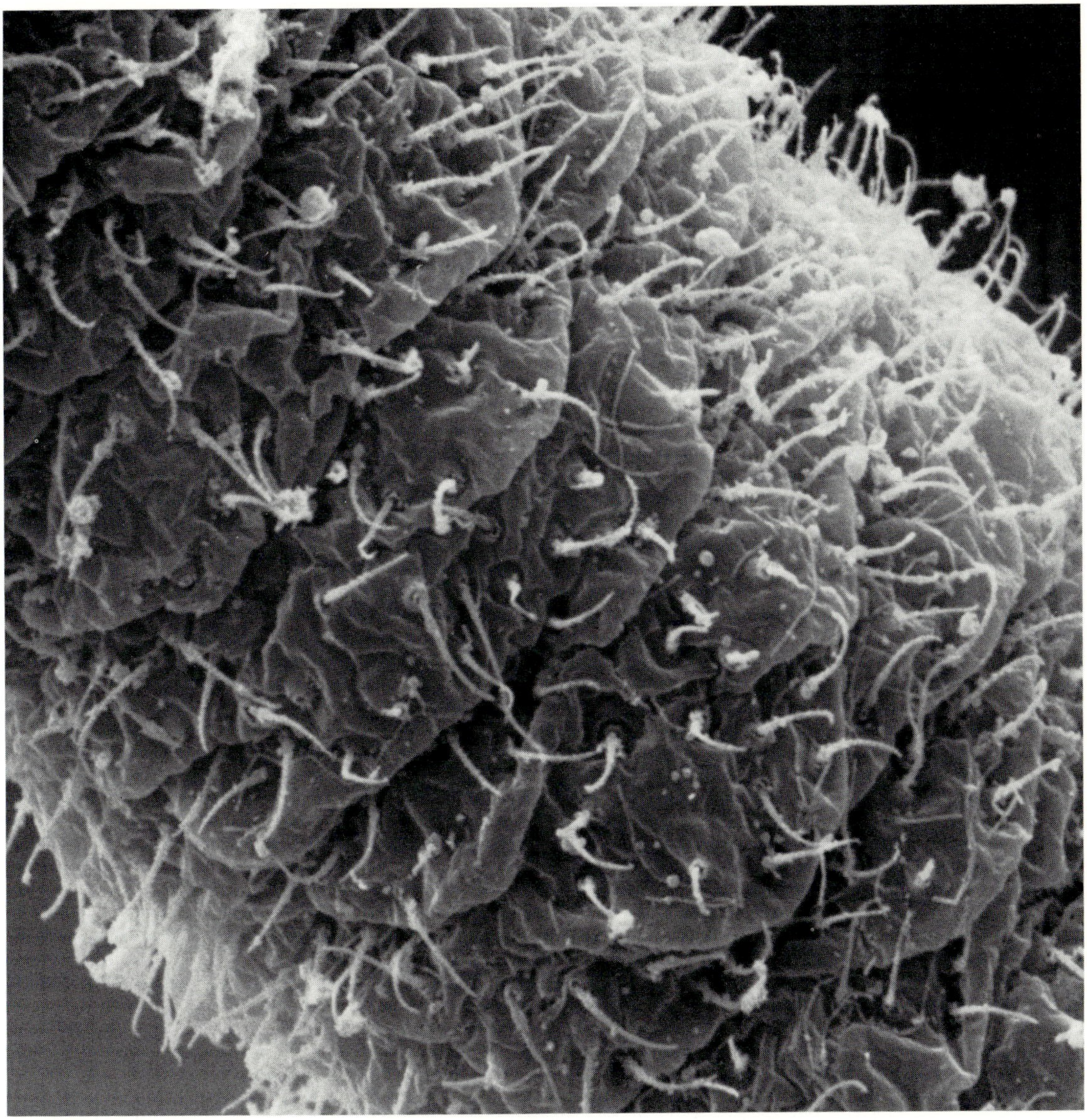

*Detail of a contracted tentacle* [x 1860]

The tentacles of coelenterates are armed with thread-cells or nematocysts. Most discharge explosively on contact with enemies or potential prey, everting a very long, hollow thread through which poison is injected. *Hydra*, like most coelenterates, has several types of nematocysts but mainly the large stenoteles are shown here. They have discharged but the threads, which are very fine, have broken off near the base. After firing, the stock of nematocysts is replenished from the interstitial cells of the column, the developing nematocysts migrating upwards through the tissues to their active sites on the tentacles.

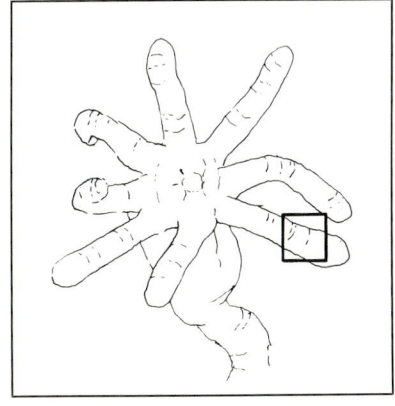

# Foraminifera

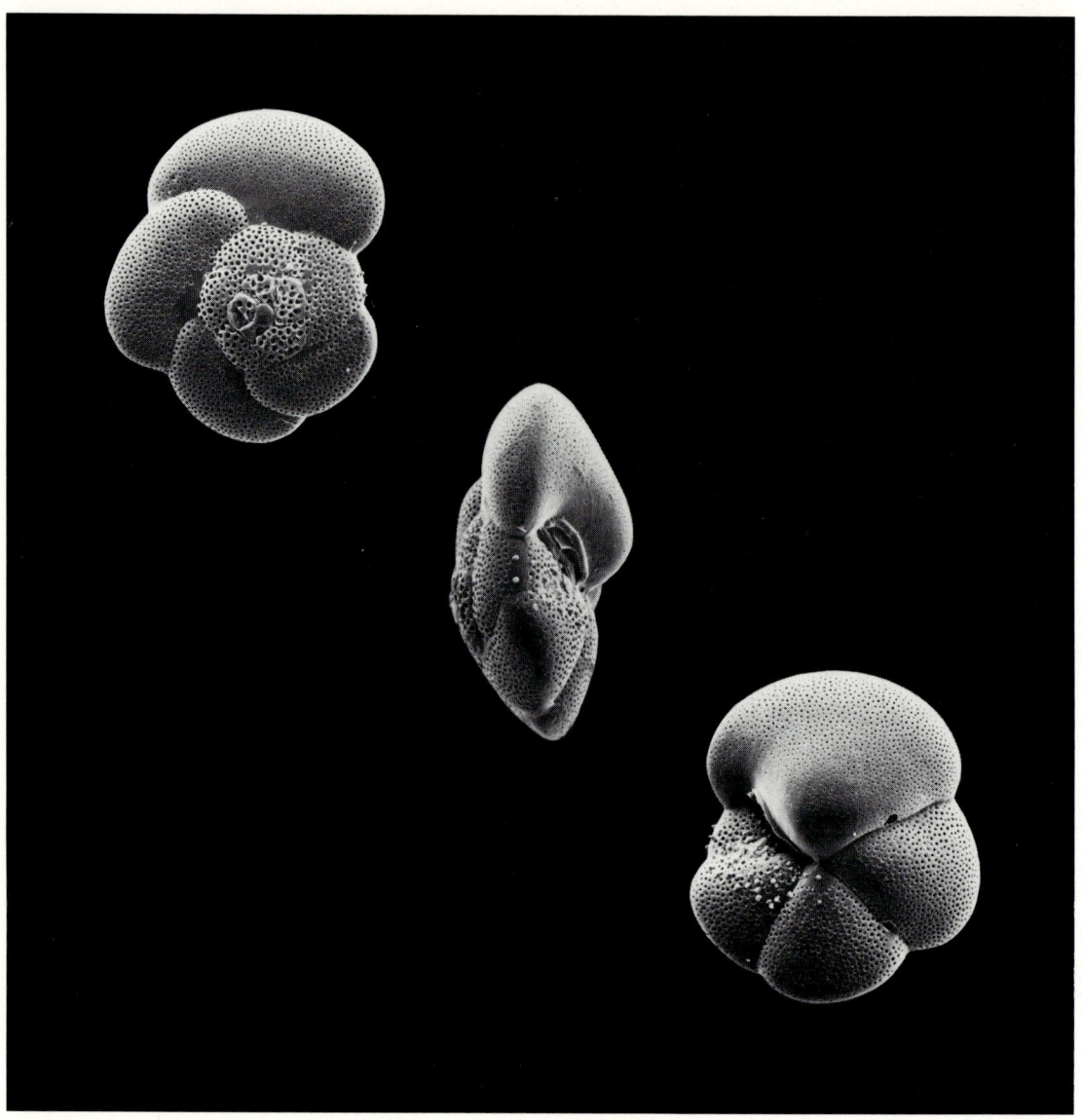

*A fossil planktonic species* Globorotalia scitula [x 95]

Foraminifera are single-celled animals (Protozoa) that inhabit marine and brackish water and secrete a shell, usually of calcium carbonate or of sand grains cemented together. Most are microscopic, this one being about half a millimetre in diameter. In life the shell contains cytoplasm which is extruded as pseudopodia through holes (apertures) and, in some cases, through fine perforations as well, and serve to catch food, mainly minute animals and algae. This illustration shows three views of a beautiful fossil species from Pliocene rocks, about 3.5 million years old in Ecuador.

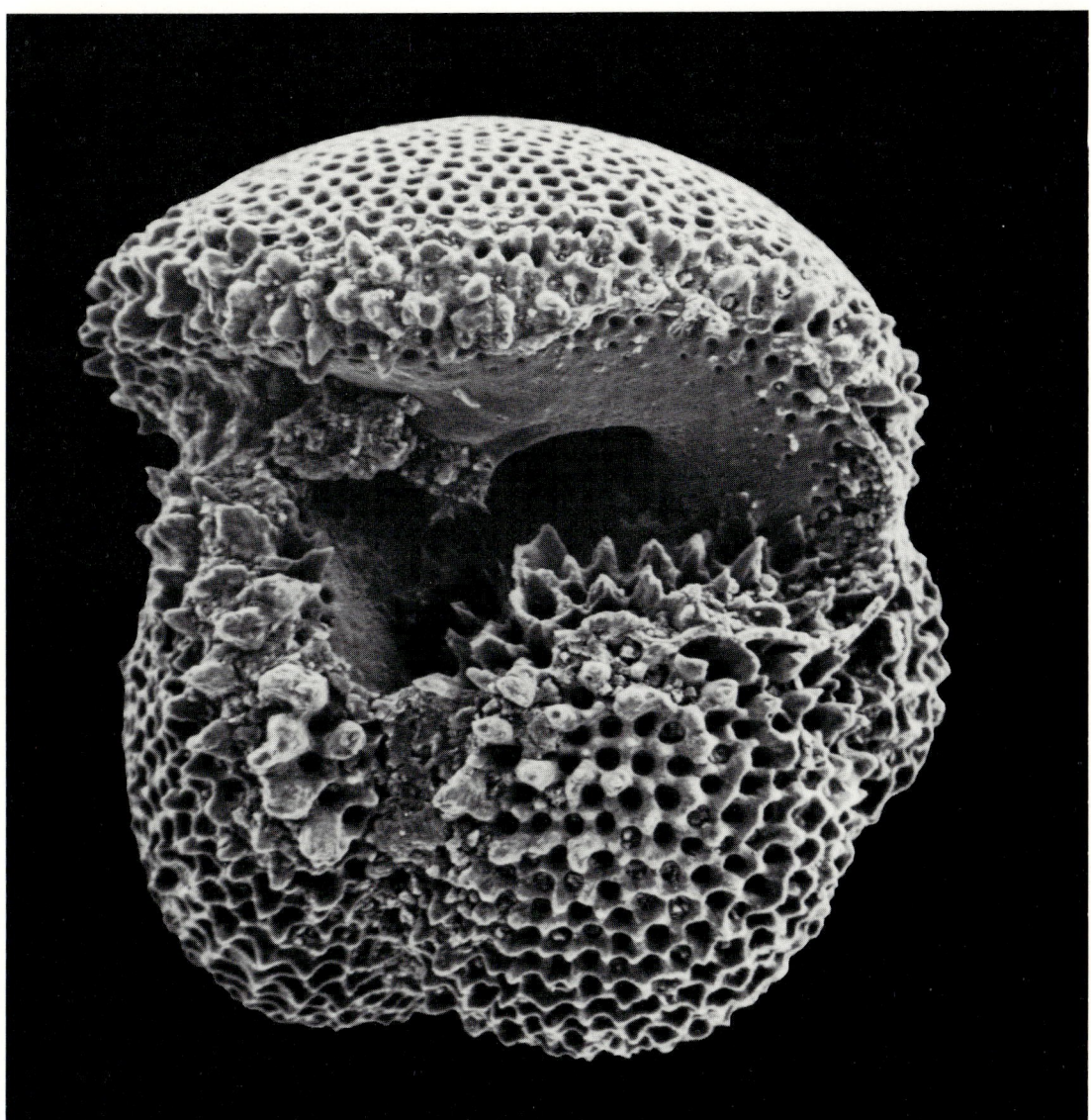

*Detail of a fossil species,* Globoquadrina dehiscens
[x 265]

This is a planktonic foraminifer from Miocene rocks
in Ecuador about 14 million years old. Some planktonic
foraminifera have long spines in life but these break
off when the dead shell becomes incorporated in the
bottom sediment of the ocean. The large central hole is
the aperture which is one of the main points through
which the pseudopodia are extruded; note also the
small perforations all over the shell.

# Foraminifera

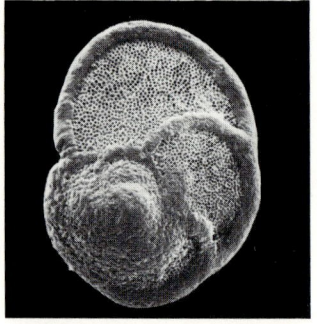
*Globorotalia tumida*[x 53]

*Orbulina universa* [x 66]

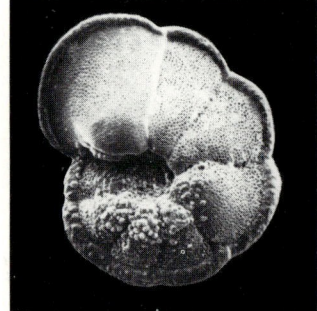
*Globorotalia multicamerata* [x 41]

*Globigerinoides tenellus* [x 145]

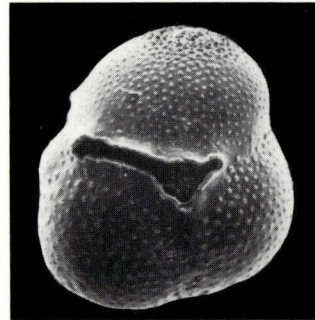
*Sphaeroidinella dehiscens* [x 67]

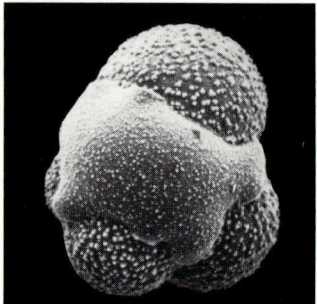
*Globigerinita glutinata* [x 250]

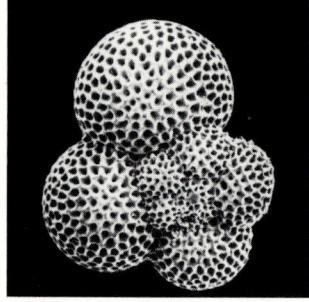
*Globorotaloides hexagona* [x 85]

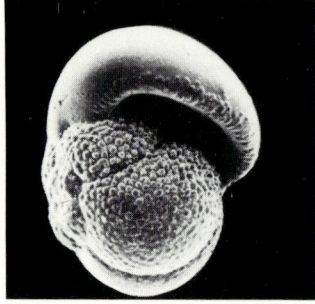
*Pulleniatina primalis* [x 75]

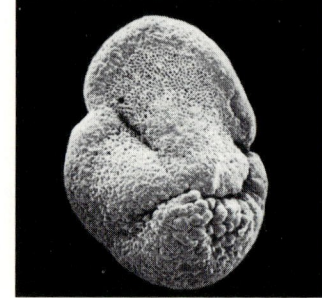
*Globorotalia tumida* [x 40]

*Diversity of form in fossil planktonic species*

Foraminifera are abundant amongst the plankton — the great community of small animals and plants in the surface waters of the sea. When they die their shells accumulate on the sea bed where they are often the dominant component of the sediment or 'ooze'. This process has been going on for countless millions of years. To the geologist, planktonic foraminifera are very useful for dating rocks derived from marine sediments, as these organisms evolved rapidly, occur in large numbers and are widely distributed. The species illustrated here are from Pliocene rocks in Ecuador (3.5 — 3 million years old).

*Fossil deep-water species*

Those foraminifera that live on the sea bed — the benthic species — are, when fossilized, useful indicators of the environmental conditions at the time the sediment was deposited. Many living species are restricted to certain types of substrate or depth of water and this information can be used, with caution, to help reconstruct the conditions under which a particular rock, containing analogous species, was formed. These examples are from Pliocene rocks in Ecuador, about 3.5 million years old. They belong to the genera: (a) *Stainforthia* [x 60], (b) *Uvigerina* [x 63], (c) *Lagena* [x 170], (d) *Rectuvigerina* [x 42], (e) *Stilostomella* [x 60].

# Ciliate

*A peritrich ciliate* [x 800]

Ciliates are complex protozoans (single-celled animals) that are abundant in most aquatic habitats. Peritrichs may be solitary or colonial, and the majority are sedentary, being attached to the substrate by a stalk. The cell-body or zooid lacks locomotory cilia, but oral cilia are present as a prominent band encircling the anterior end of the cell. This species, *Campanella umbellaria*, is unusual in that oral cilia encircle the apical end of the cell about four and a half times before turning downwards into the mouth. Peritrich ciliates are important indicators of water quality. As they feed by filtering bacteria from the water, they are used in waste water treatment.

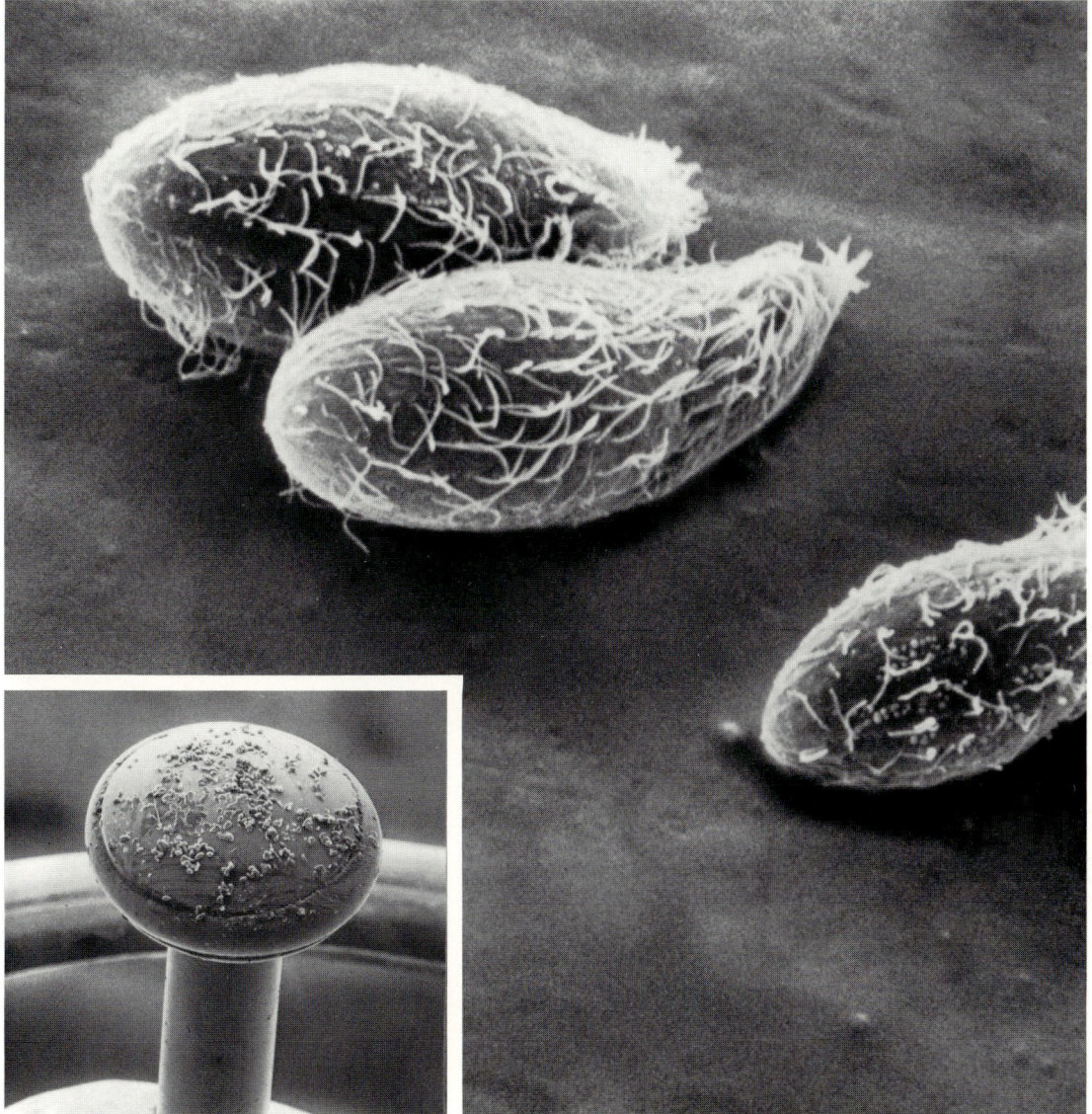

*Hymenostome ciliates on a pin-head* [x 1520 and 19]

The simple nutritional requirements of this species, *Tetrahymena elliotti*, together with its high growth rate have contributed to its popularity in the field of protozoan research. Probably more is known about *Tetrahymena* than any other ciliate and it has been employed in many studies including ultra-structural research, genetics, physiology and biochemistry. The body is ovoid, with a mouth, bounded by three membranelles, at the anterior pole. The cell surface is equipped with rows of hair-like cilia that beat in a coordinated rhythm to propel the organism forwards. *Tetrahymena* is a free-living freshwater ciliate.

# Plants

Whereas in animals most cells have exceedingly thin, membranous walls and any rigid skeletal substance is formed outside the cells, in plants the majority of cells develop a characteristic shape by the deposition of strengthening material, especially cellulose, within the cell wall. As a result most plant surfaces reveal a great amount of detailed microstructure on close inspection and scanning electron microscopy can also be used to study internal tissues like wood. Plants are well represented amongst the microscopic forms of life and diatoms in particular show a wealth of detail on a scale that reaches and far exceeds the limits of light microscopes.

# Dandelion

*Vertical section of an unripe fruiting head* [x 27]

The flower head of a dandelion, *Taraxacum officinale*, as in other composite flowers, consists of a large number of separate flowers or florets arranged on a flat-topped receptacle. Each floret produces a bottle-shaped, single-seeded fruit or achene, topped by a tuft of hairs, the pappus. When the fruit is ripe the pappus beak grows to form a parachute that allows the wind to carry the fruit and the ripe seed inside away from the parent plant.

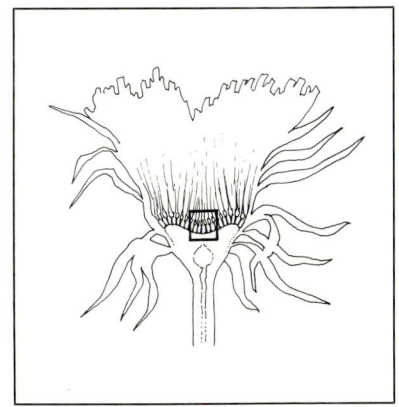

# Dandelion

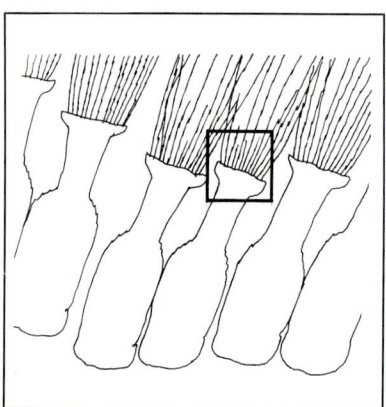

*Attachment of the pappus of 'parachute hairs' to the unripe fruit [x 136]*

When the fruit is ripe, the beak with the pappus attached elongates, making the long 'stem' of the parachute.

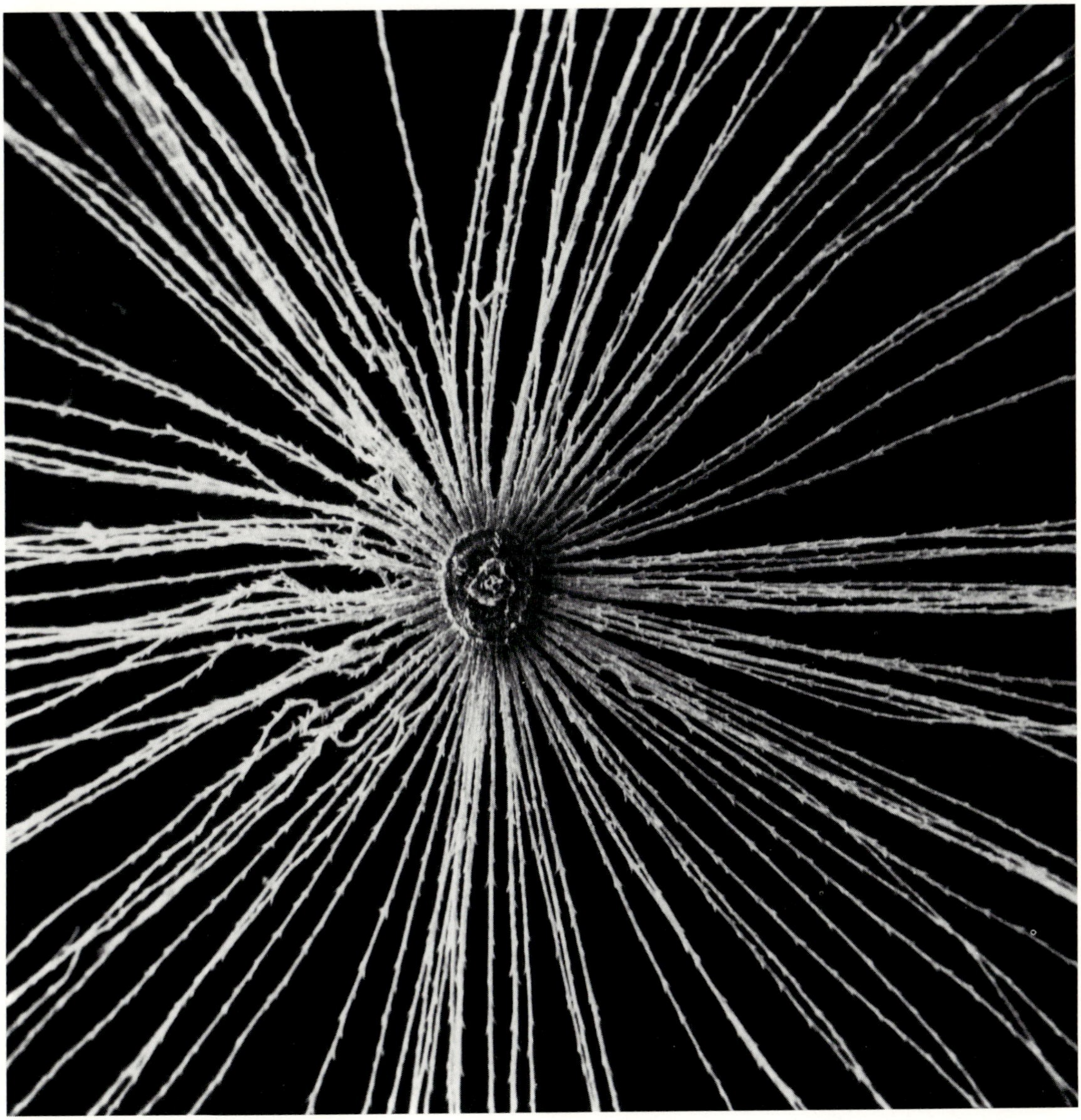

*The expanded pappus or parachute viewed from above
[x 27]*

After fertilization, when the seed has ripened in the
fruit, the stamens, stigma and petals wither, the beak
grows away from the fruit and the pappus opens out
into an umbrella or parachute shape. When conditions
of temperature and humidity are right the whole fruit is
blown away from the plant by the wind.

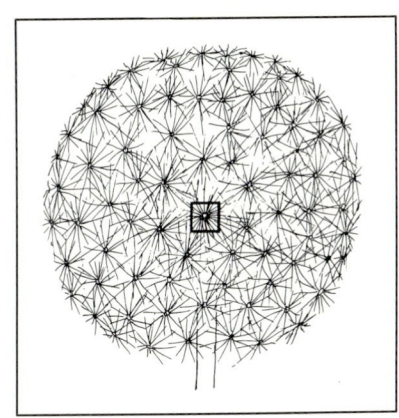

# Dandelion

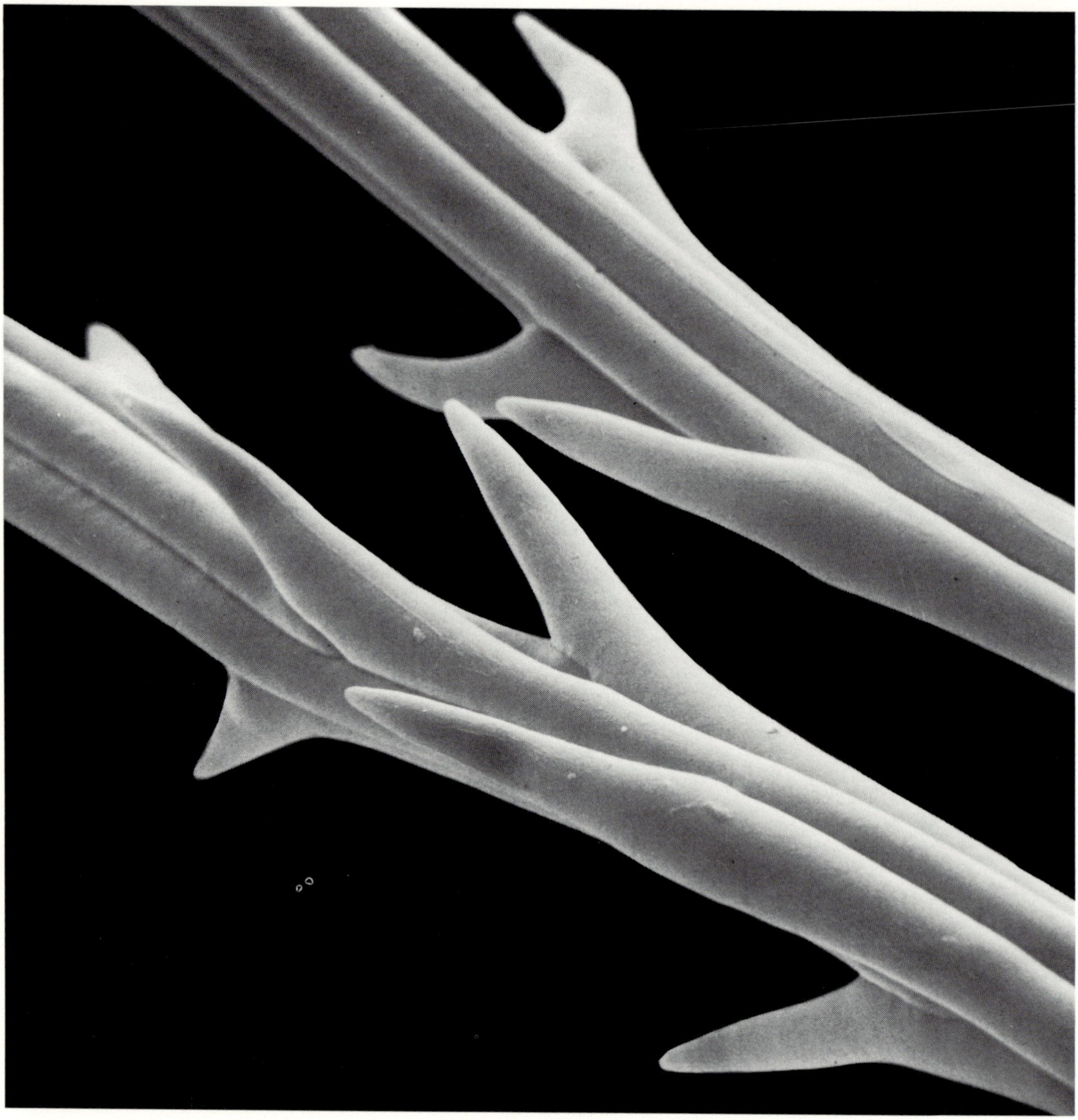

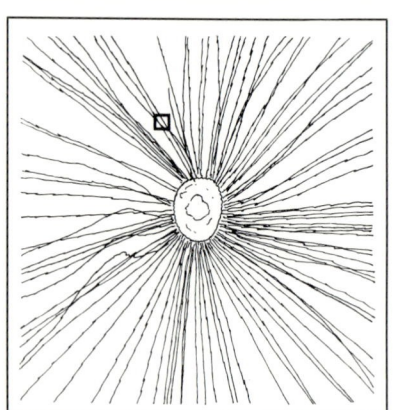

*Parts of two pappus hairs* [x 1500]

The function of the barbs is obscure since many related species lack them. They probably arise through the fusion of hairs. Similar but longer barbs are found on the fruit.

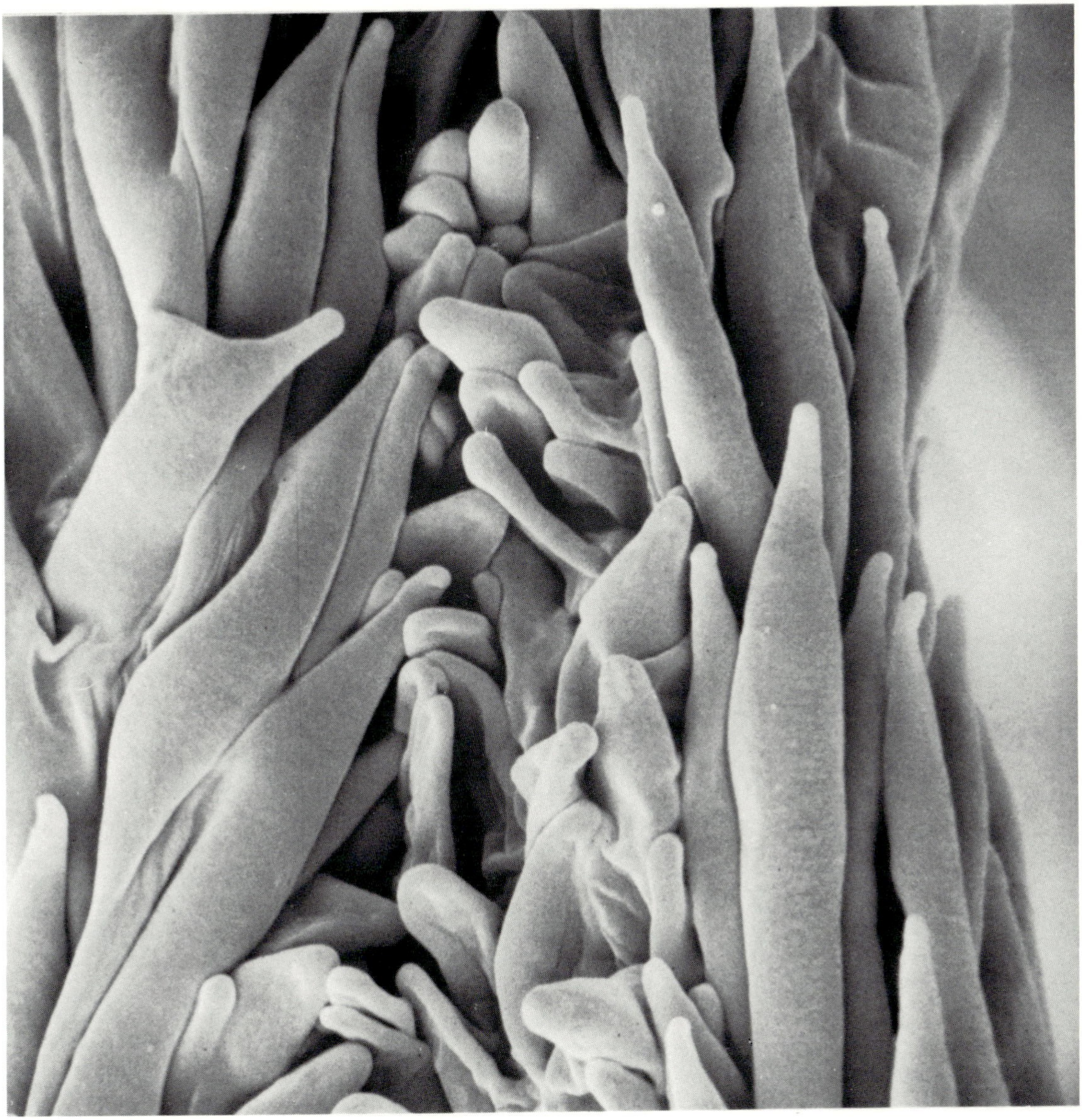

# Dandelion

*Part of a stigma* [x 1380]

The stigma is the receptive area of the female part of
the plant to which the male pollen adheres. Once on the
stigma, pollen grains release characteristic proteins.
The surface of the stigma is covered by complex
chemical recognition substances that stimulate
germination of compatible pollen. Successful
pollination depends upon mutual chemical recognition
between stigma and pollen grains.

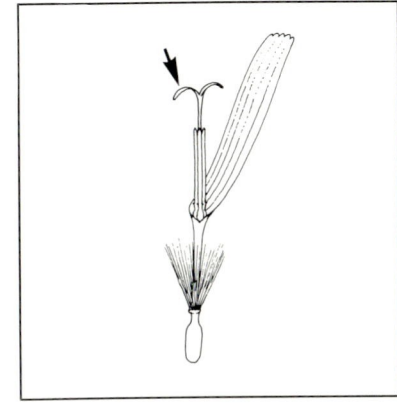

# Dandelion

*A grain of pollen* [x 2760]

The outermost layers of the grain form an elaborate system of spiny ridges. This arrangement is characteristic of the pollen of the chicory tribe of which the dandelion is a member. There are three pores, two of which are visible at the lower left and right of the grain, showing the inner layers of the wall bulging through the pore areas. (The bulging is an artefact caused by the vacuum in the microscope.) These are the areas from which the pollen tube will grow.

*Surface of a petal* [x 1380]

The deep corrugations are a characteristic feature of many flower petals. This one is from a common daisy, *Bellis perennis*. The curved cells act as lenses that are correlated with the distribution of pigment and project light to precise positions in the cells. The colour and brilliance of the petals are of great importance in attracting pollinating insects.

# Daisy

*Surface of an anther* [x 1380]

Stamens develop in a similar way to petals as can be seen especially in some 'double' garden varieties where various intermediate conditions are found. It is therefore not surprising that the head or anther of a stamen shows some similarity in surface pattern to a petal.

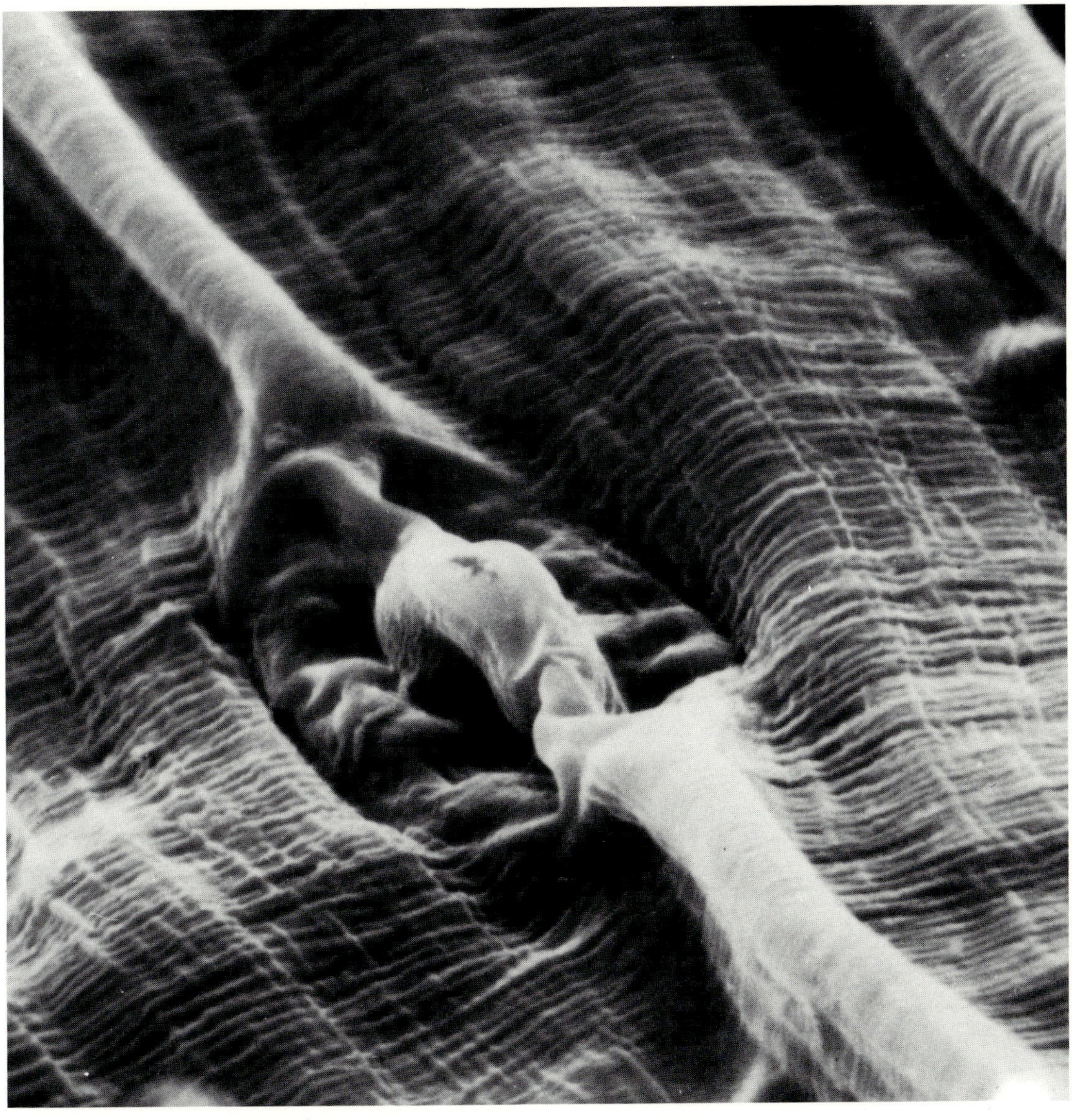

*Stoma on the surface of a stem* [x 3040]

Most of the outer surface of a plant is covered by a thin waterproof membrane or cuticle. Movement of water and gasses takes place through pores or stomata found on the leaves and stem. These open and close mainly in reponse to light, being generally open by day and closed at night. The water taken in by the plant's roots eventually passes out of the stomata in the form of water vapour or even as drops of water.

# Milkwort

*Seed* [x 30 and x 50]

Seeds show an enormous diversity of surface features, the significance of which is rarely apparent. The survival value of structures such as the spines and protuberances on these seeds of milkwort, a species of *Polygala*, is currently being investigated.

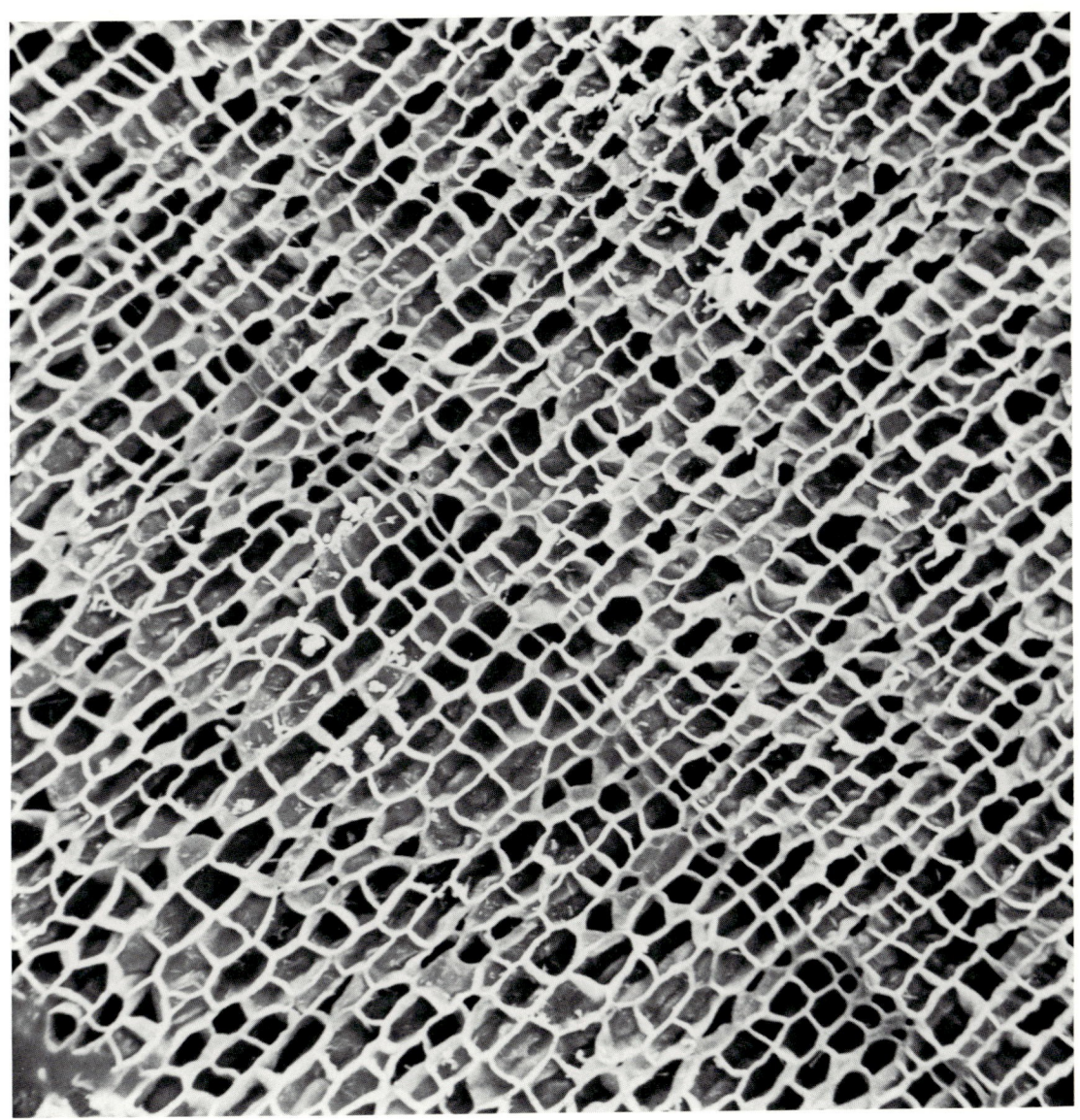

*Transverse section of cork from the cork oak* [x 138]

Cork is the basic constituent of all tree bark and consists of cells that are proliferated from the outside of the living layer of the bark, the cork cambium. These cells soon die and in most trees contain hard waste substances, but those of the cork oak, *Quercus suber*, remain resilient. Bark for commercial purposes is removed from the tree after about twenty years growth; this first cork is of little value but subsequent strippings at intervals of ten years yield a high quality product.

# Cork

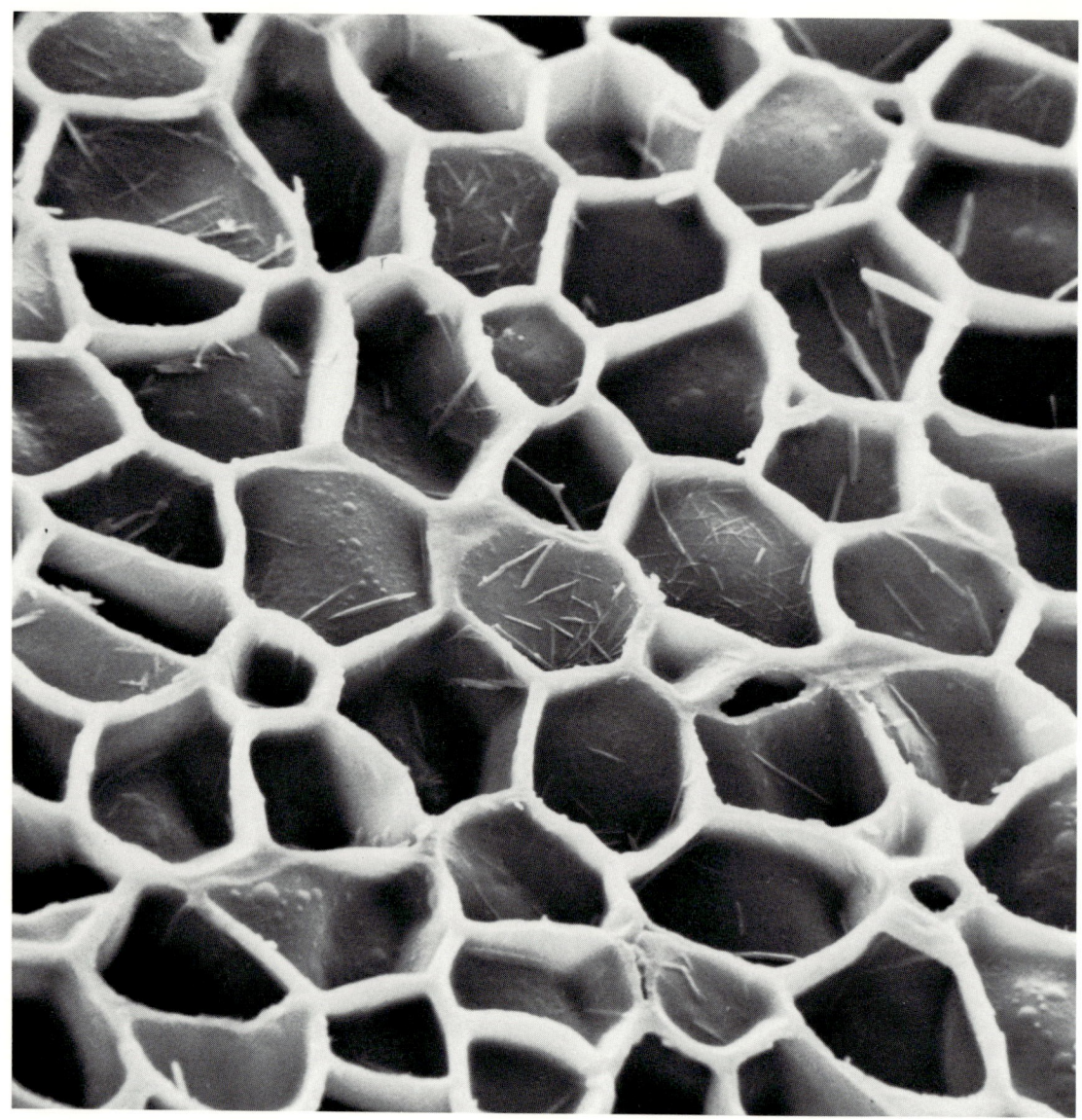

*Detail of cells* [x 690]

The cell walls of cork contain a fat-like substance called suberin which gives it the properties of water resistance and impermeability, thus making it an ideal material for stoppering bottles containing liquids and for making various insulating materials and buoyancy aids.

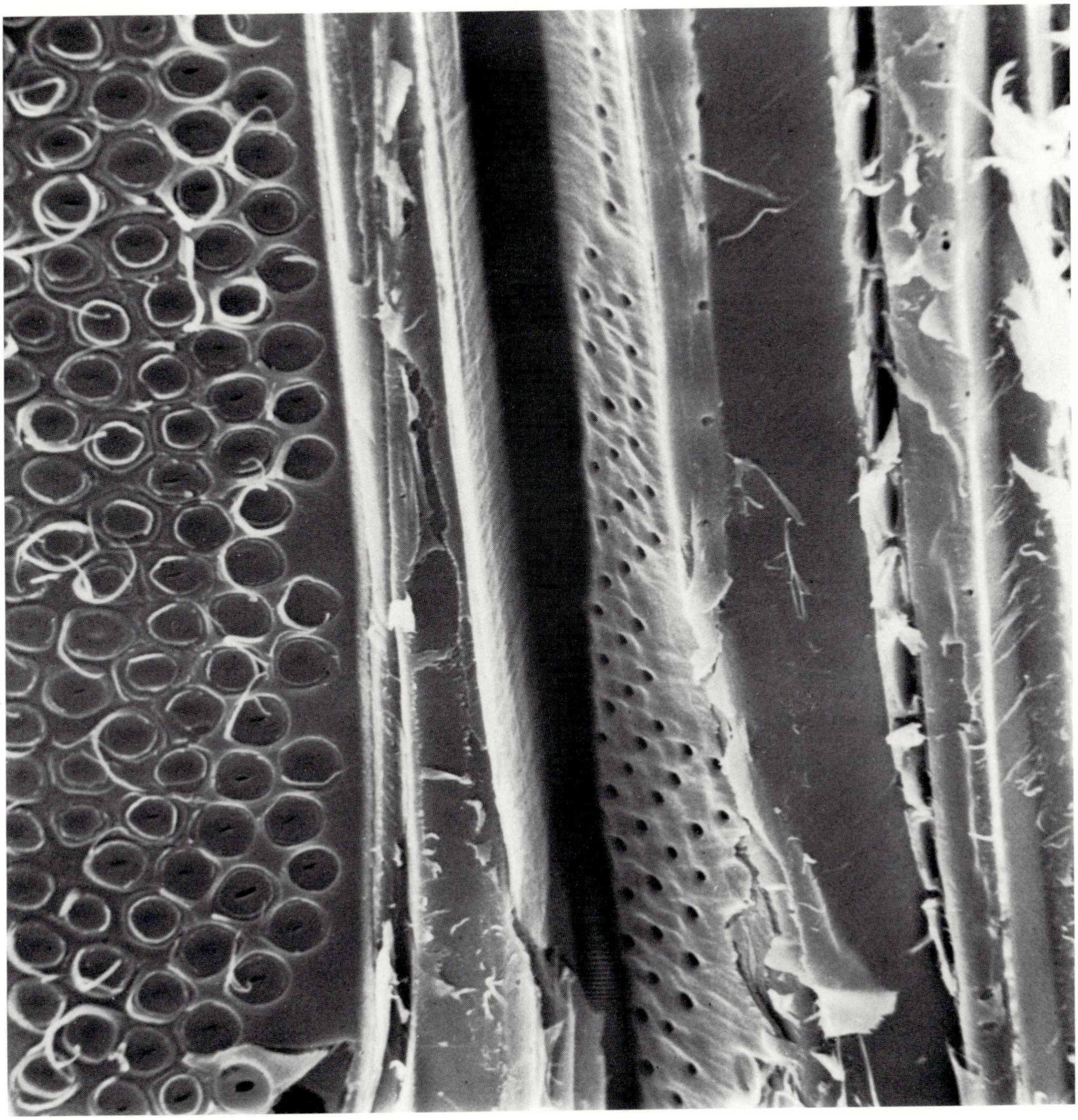

*Fractured surface of a matchstick* [x 690]

The aspen, *Populus tremula*, grows quickly and
produces a soft wood that is of little use for
construction but ideal for making matches. Wood is a
complex structure. On the left two water-conducting
'vessels' have been torn apart revealing the pores in
the cell wall through which they interconnect. In the
centre the wall of a similar vessel is seen from the
inside of the cell showing the small apertures of
the pores.

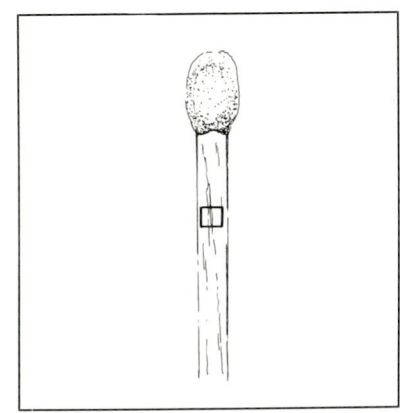

# Aspen

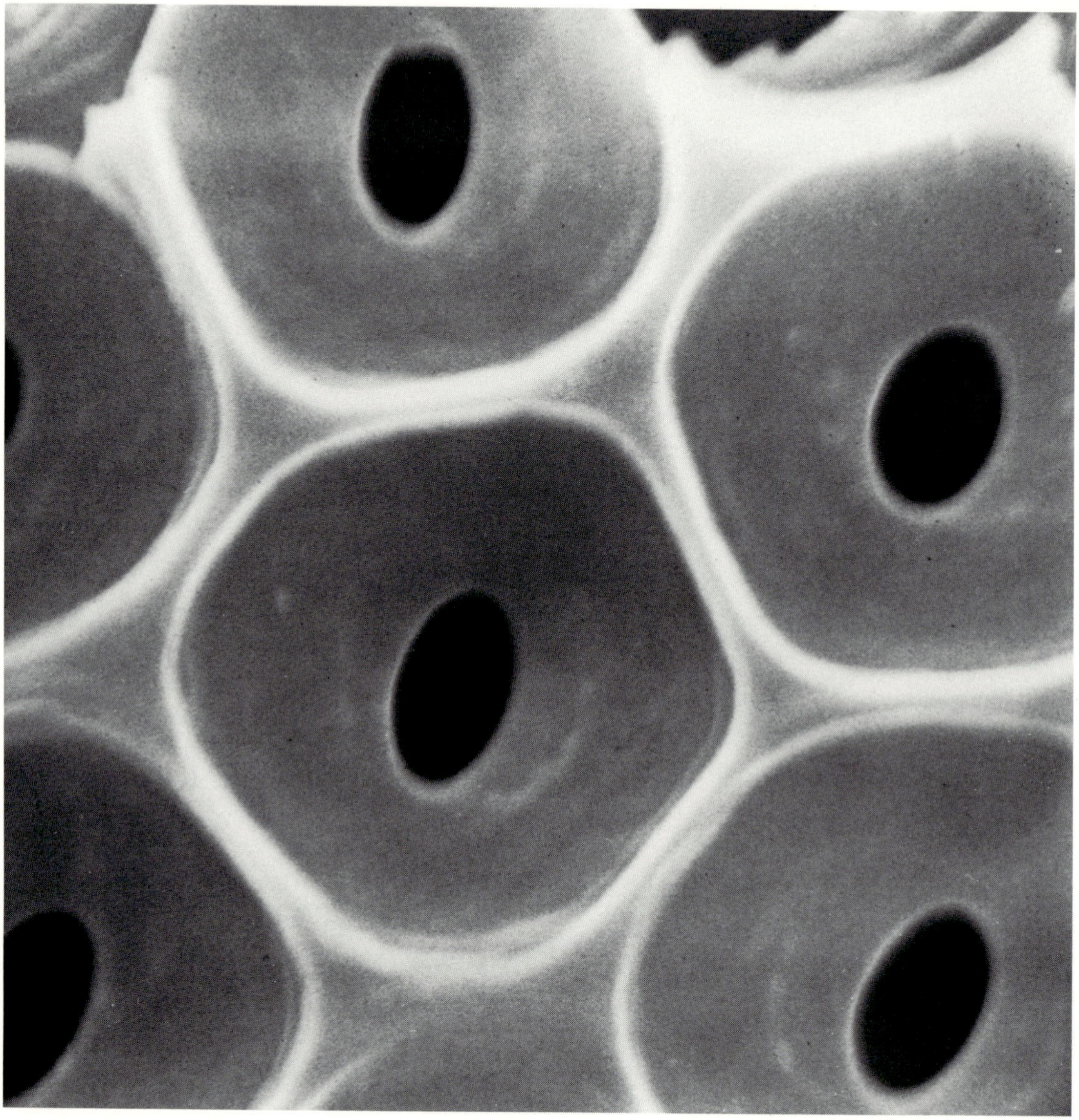

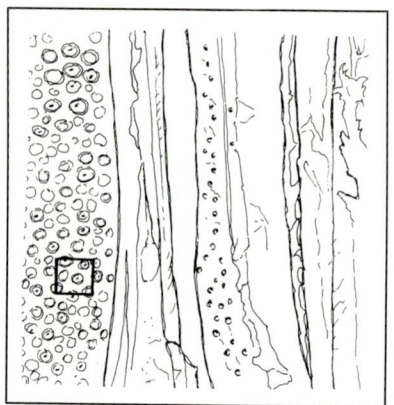

*Detail of conducting vessel* [x 6900]

A small part of the cell wall separating two water-conducting vessels in the wood. These 'bordered pits' are complex structures allowing passage of water and solutes from one vessel to its neighbour. In this specimen two vessels have been separated revealing the insides of the circular pits, which are contained within the thickness of the cell walls, and the smaller openings by which the pits connect with the main cavity of the vessel. The membrane that divides each pit in the central plane of the wall has been removed.

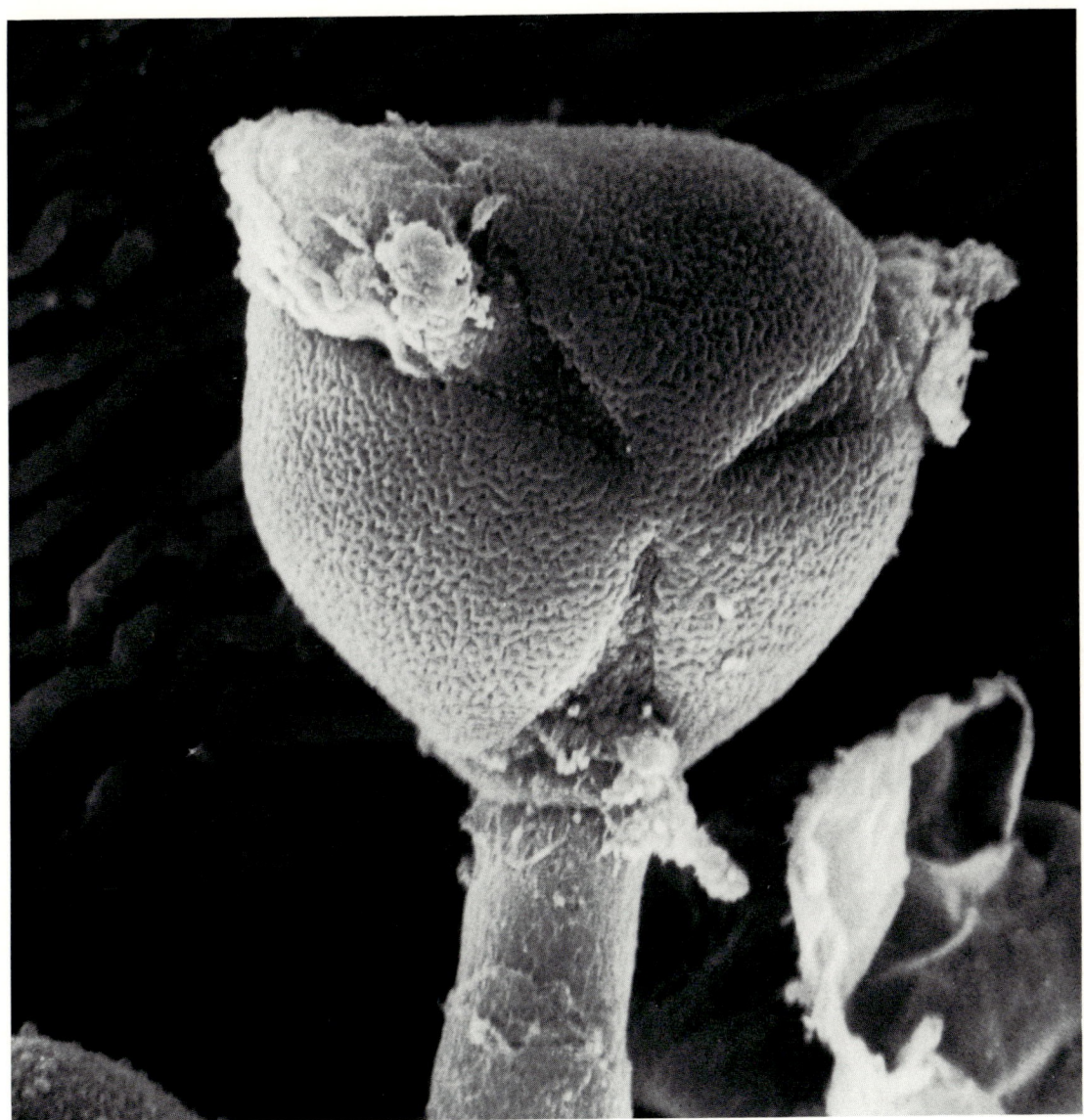

*A pollen grain of* Paeonia delavayi *with a pollen tube*
[x2760]

Although pollen grains effect fertilization of the female
plant they are not strictly analagous to the spermatozoa
of animals. Each grain is a two-celled spore. The
vegetative cell germinates on arrival on the (female)
stigma of the flower by growing a tube that pushes
down inside the stigma to reach an ovule. The nucleus
of the generative cell divides and one nucleus travels
down the pollen tube into the ovule to combine with the
female nucleus. This photograph shows a pollen tube
(bottom) that has grown out from one of the three
openings in the surface of the grain.

# Pollen

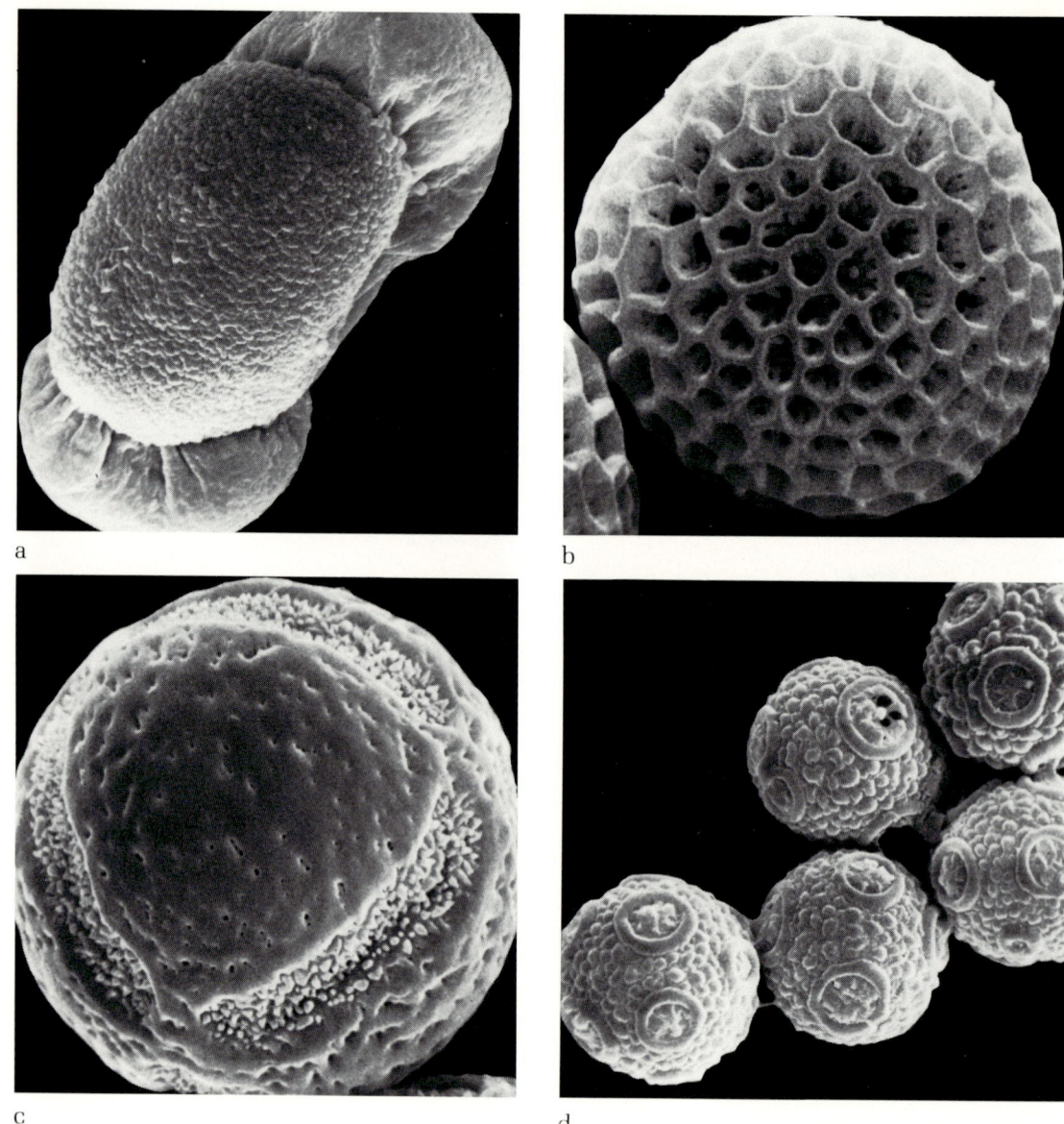

a

b

c

d

*Variety in structure*

The outer coat of a pollen grain is very resistant and retains its characteristic features when fossilized. The great variation of pattern seen in pollen grains can therefore be used to analyse fossil deposits, such as ice-age peat, to determine the identities and relative abundance of the dominant species of plant growing at the time the deposit was formed. The four diverse examples of living pollen shown here are (a) Scots pine, *Pinus sylvestris* [x 950] — note the balloon-like air-sacs that allow dispersion by the wind; (b) stock, *Matthiola incana* [x 2680]; (c) bleeding heart, *Dicentra spectabilis* [x 2290]; (d) fumitory, *Fumaria occidentalis* [x 625].

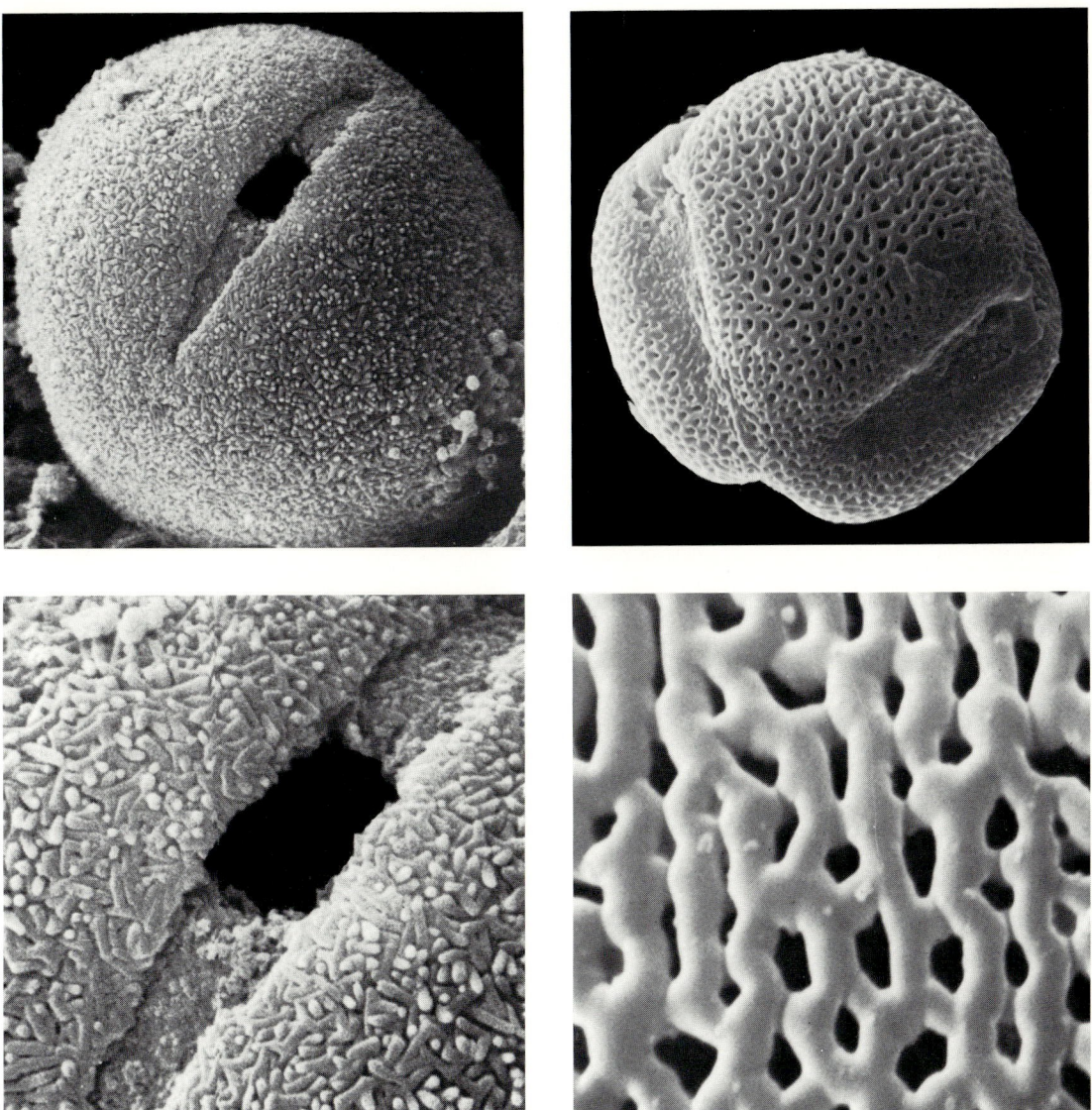

*Detail of two grains*

Left: beech, *Fagus sylvatica* [x 1660 and 4100],
showing one of the apertures through which the pollen
tube grows. These apertures change in size according
to atmospheric conditions, so that the entry and loss of
water and gasses are controlled. Right: felwort,
*Gentianella amarella* [x 1640 and 8210].

# Pollen

*Surface complexity*

The significance of the complexity of pollen grains is little understood. Adaptations for transport by different mechanisms such as wind, insects etc. and for carrying the proteins which are recognized by the stigma of the correct species of plant are probably important factors. These examples are (above): hollyhock, *Althaea rosea* [both x 1500] showing (left) outside of grain and (right) a cut section showing the pores on the inside of the wall; (below): *Aulojusticia linifolia*, an African shrub [x 1025 and 3790].

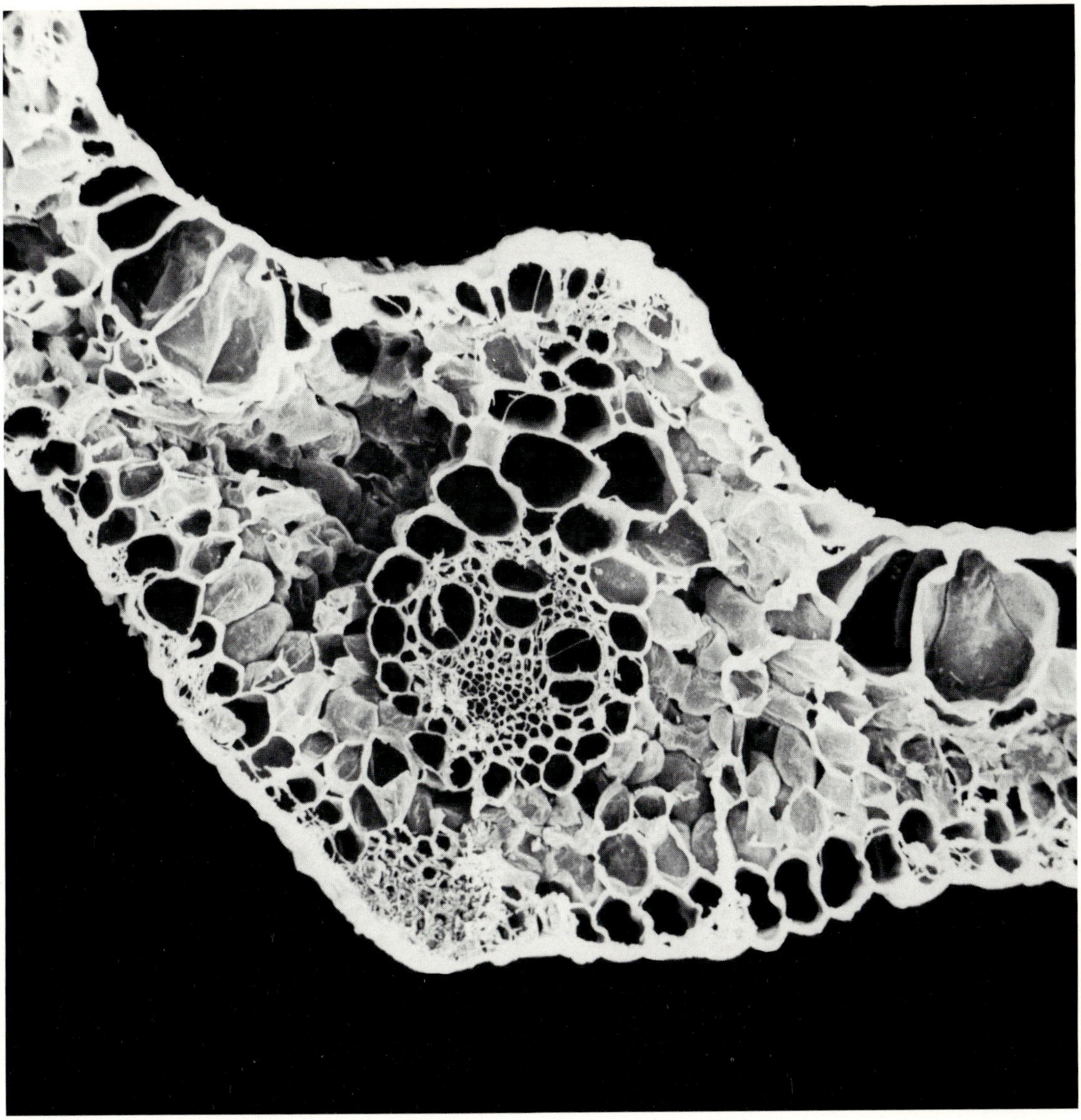

*Section of a blade of grass from the stomach of a cow*
[x 300]

The digestive enzymes produced by a mammal's alimentary canal cannot break down the cellulose that encloses most plant cells. In this thin section of a blade of rye grass, *Lolium perenne*, some cells have been cut open, others remain intact.

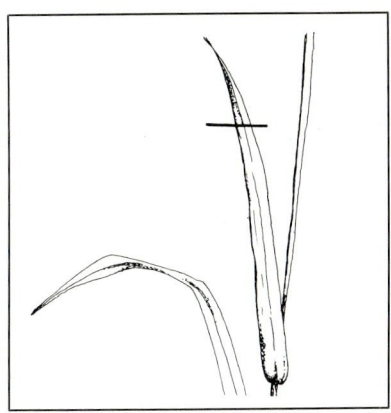

# Digestion of grass

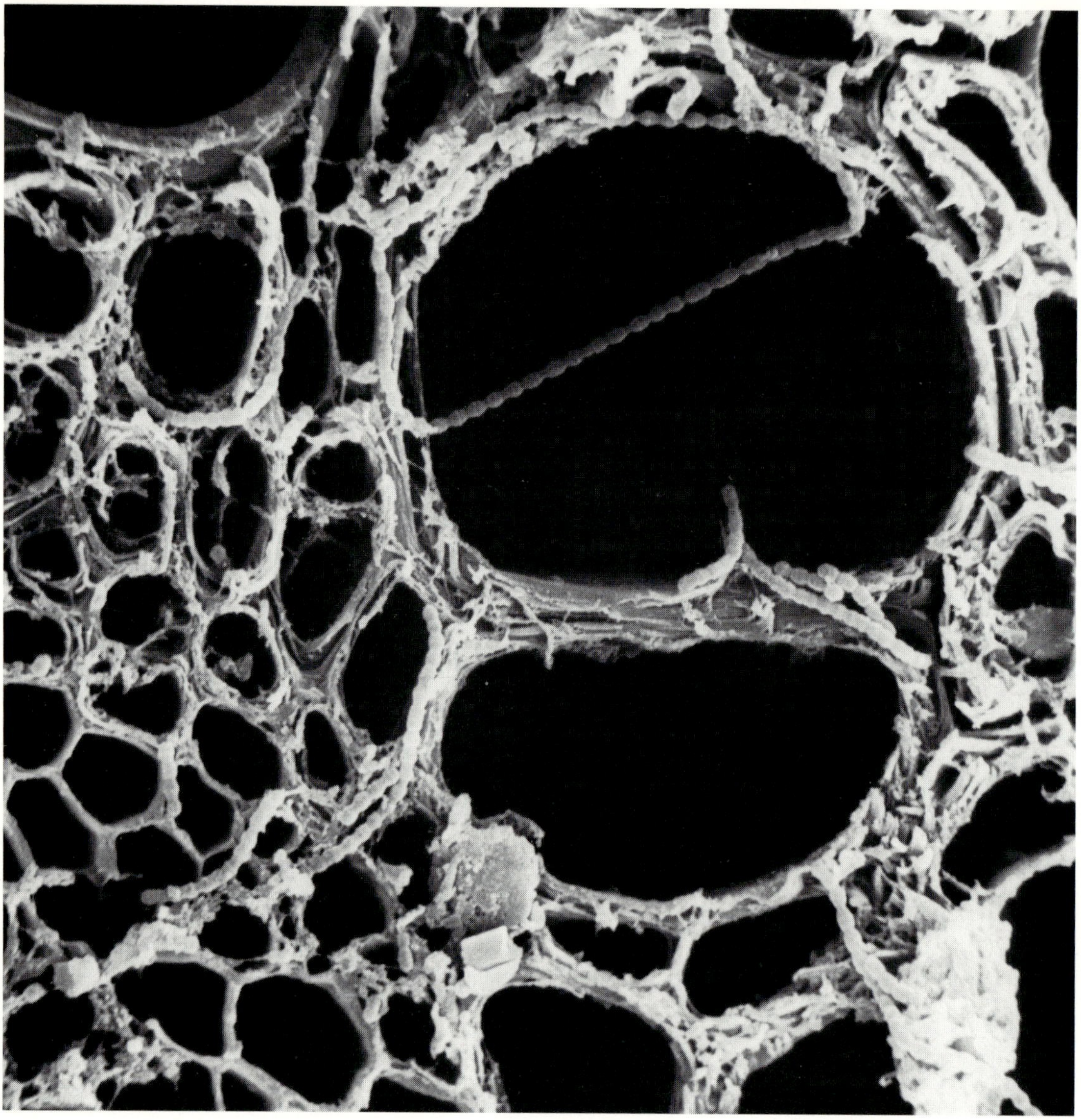

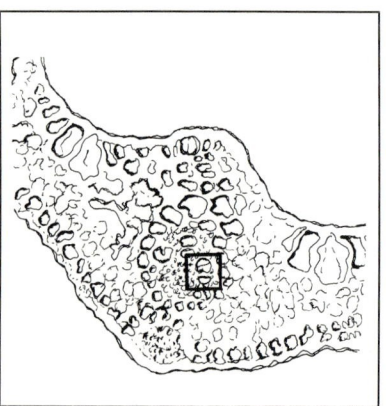

*Bacteria in cells of grass* [x 3040]

In the stomach of a cow the cellulose of the grass cells is broken down by specialized bacteria. This higher magnification of the previous subject shows one of these, *Ruminococcus flavefaciens,* in the form of long chains of cells, mostly adhering to the fractured edges of the cell walls. The bacteria can only gain access to the cellulose if the cell wall is broken, hence the importance of chewing the cud.

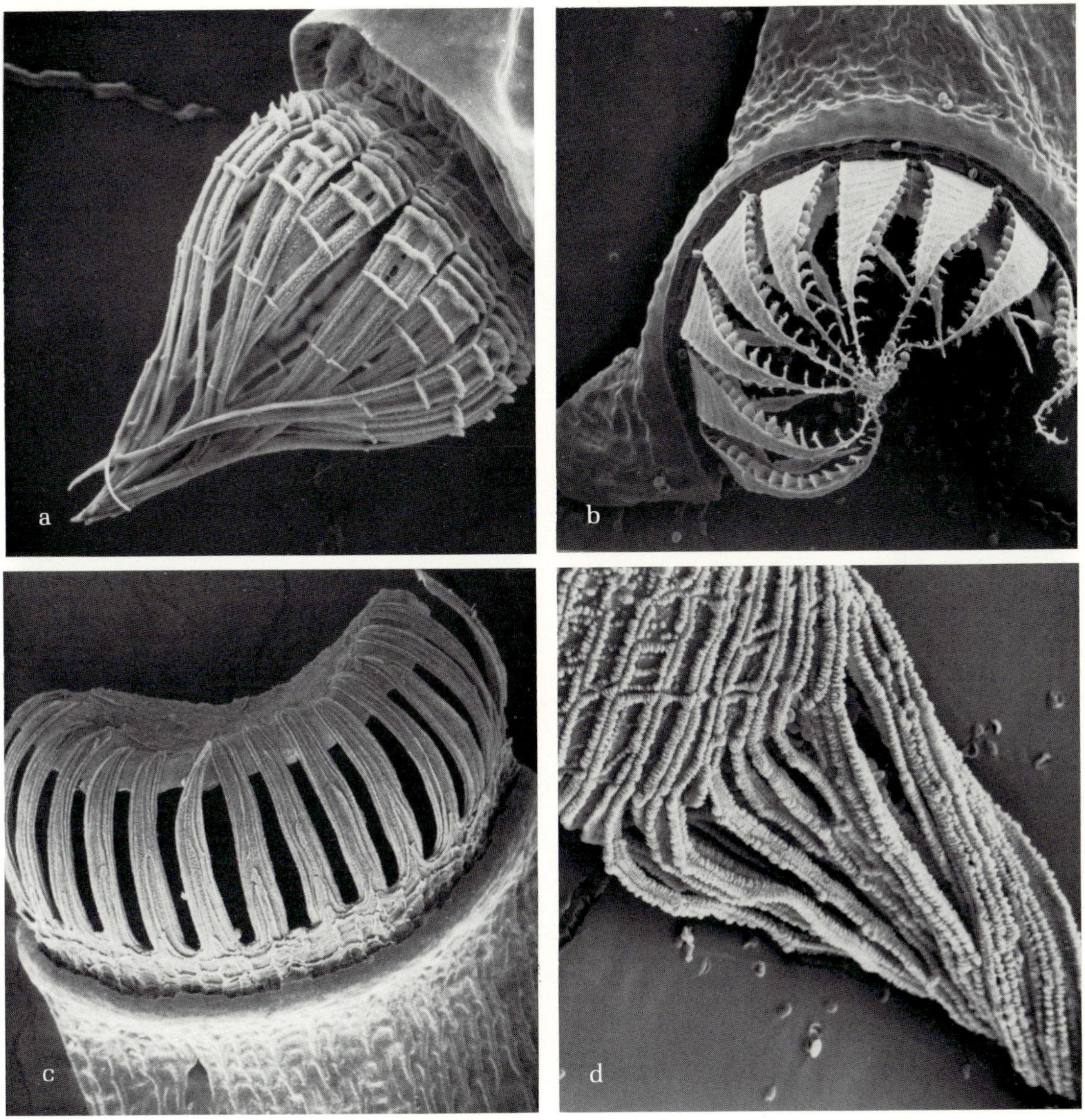

a

b

c

d

## Variation in spore-capsules

The fertilized egg-cell of a moss grows, still attached to the parent plant, into a sporophyte, consisting of a stalk and a spore-capsule. The opening of the capsule is surrounded by teeth which regulate the release of spores by reacting to humidity and show a great variety of form in different species. (a) *Ceratodon purpureus*: a single ring of 16 teeth [x 500], (b) *Funaria hygrometrica*: a double ring of 16 pairs of teeth [x 80], (c) *Atrichum undulatum*: 32 immobile teeth connected to a central ring [x 70], (d) *Tortula subulata*: a spirally twisted brush of very long teeth [x 90].

# Mould

*Spore production in a species of* Aspergillus [x 1380]

Moulds are fungi and like most fungi the main part of the plant consists of a multitude of fine, colourless, branching threads dispersed throughout the substance from which it derives its nourishment. If a mould is conspicuous it is because of the production of spore-producing organs like this, just as the underground ramifications of other fungi only become apparent when mushrooms or toadstools emerge. *Aspergillus* is common on bread, jam, old leather etc. and is closely related to *Penicillium,* the source of penicillin.

*A mixture of fossil species* [x 950]

Diatoms are microscopic, single-celled, aquatic plants
containing yellow and brown pigments in addition to
chlorophyll. The cell wall is impregnated with silica,
forming a rigid, box-like structure usually with complex
patterns of pores. Some sediments are formed almost
entirely of diatom shells, which retain their pattern on
fossilization, and such 'diatomaceous earths' or
'Kieselguhr' are used in the manufacture of filters,
polishes, insulation and many other products. This
sample is from a marine deposit of Lower Eocene age
(about 55 million years old) in the USSR.

# Diatom

*Entire shell* [x 2730]

The shell (frustule) of a diatom consists of two valves, usually of an elaborate structure as in this species, joined together by several simple overlapping hoop-like structures called the girdle. When the cell divides, each daughter cell retains one valve and half of the girdle of the parent cell and grows one new valve and half a girdle. This species, *Campylodiscus fastuosus,* lives on the bottom of shallow seas.

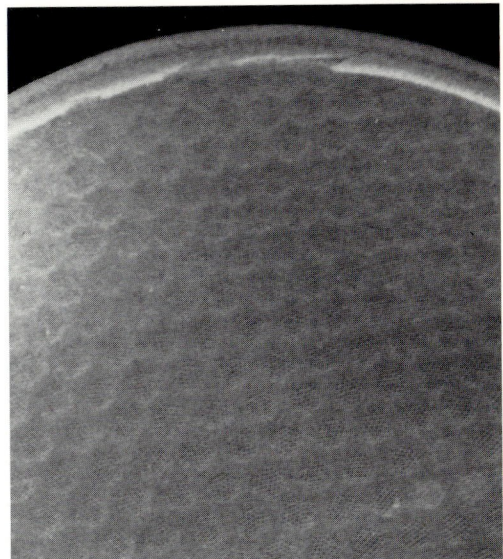
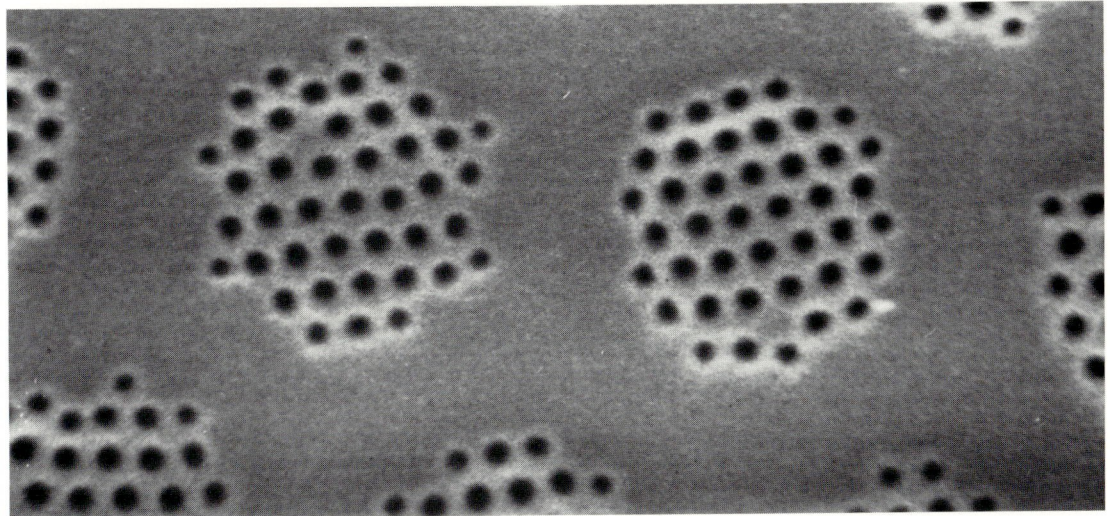

*Detail of perforations* [x 80, 780 and 14 200]

The electron microscope clearly reveals details, such as
these clusters of pores on the inner surface of a valve,
that are scarcely detectable with a light microscope.
These successive enlargements are of a fossil species
from Lower Eocene deposits in the USSR.

# Diatom

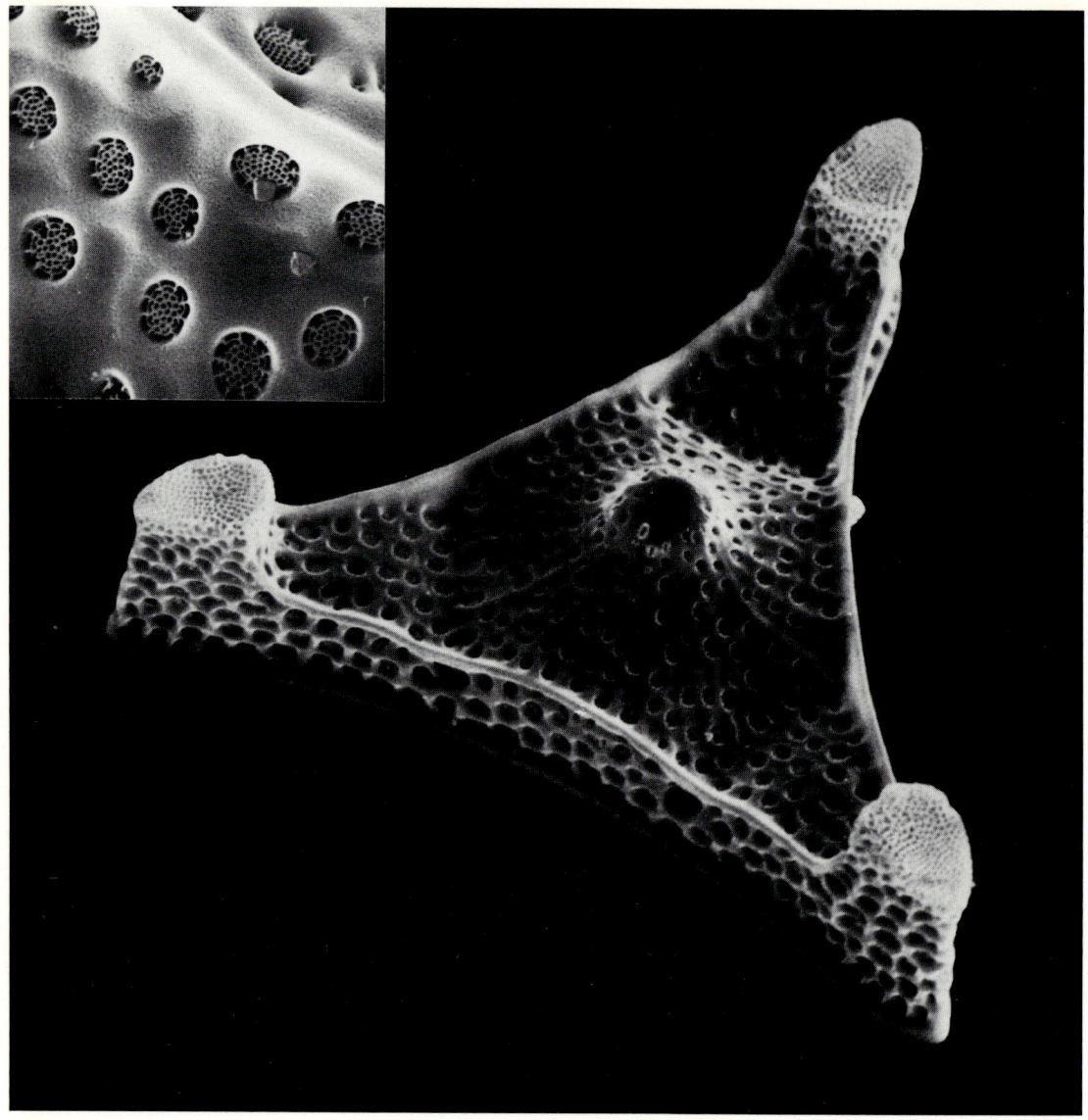

*A fossil species of Eocene age* [x 800, 5200]

Fossil diatoms usually consist of separate valves, as here, rather than whole frustules, but the intricate detail within each pore is still preserved. Living species related to this one live attached to rocks etc. by jelly-like pads that are extruded through the raised knobs at the corners. This is a species of *Triceratium* from an Eocene deposit in the South Atlantic.

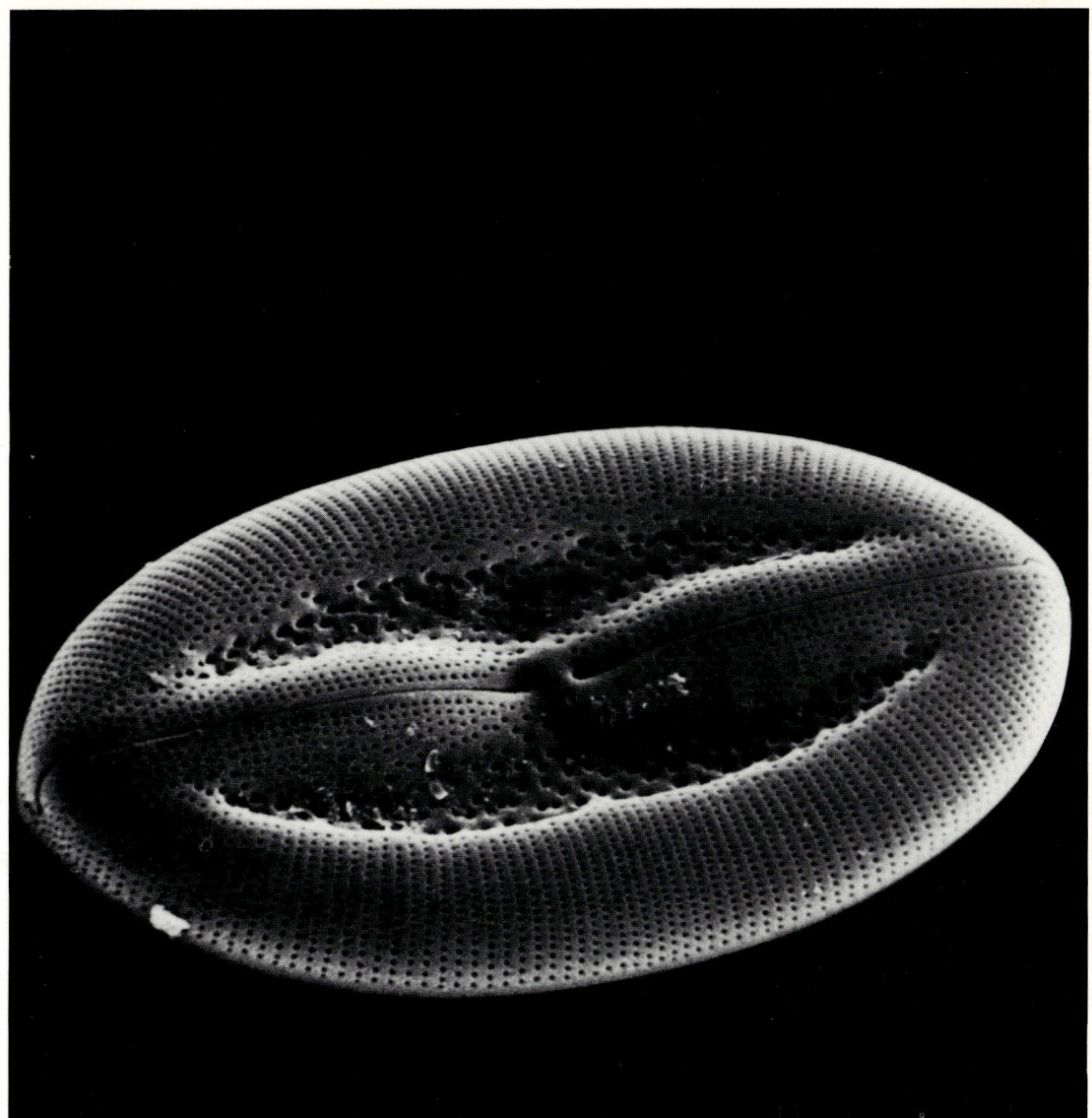

*A living species of pennate diatom* [x 1160]

Some diatoms are able to move by creeping over the surface on which they live. These species all have a long slit — the raphe — divided at the centre, in the bilaterally symmetrical silica shell. This species, *Navicula praetexta*, lives on the bottom of shallow seas; similar species are common in fresh water.

# Diatom

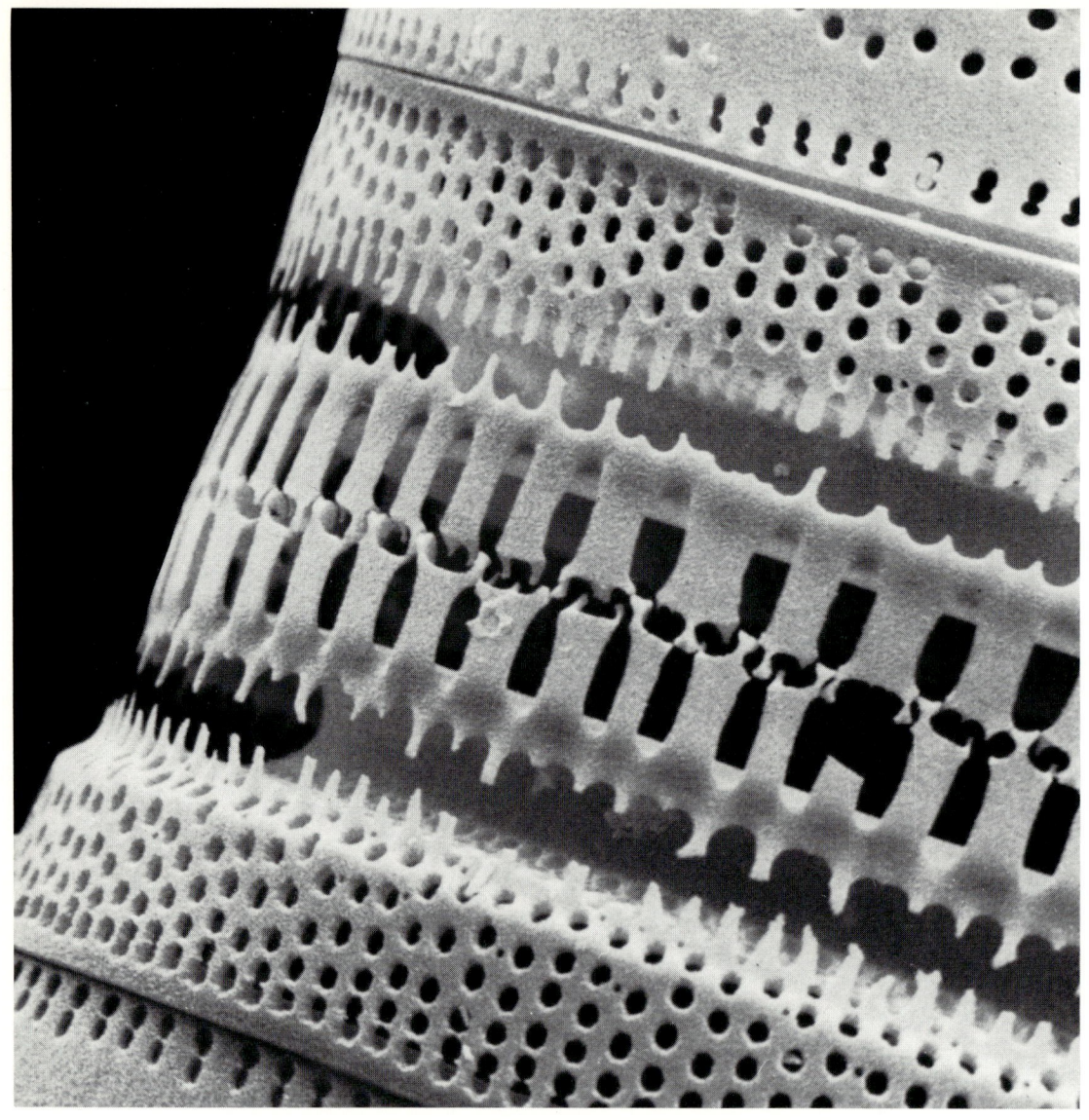

*A fossil, colonial species* [x 1660]

Some diatoms form colonies, with the individual cells linked to form a chain. This photograph shows the elaborate nature of the structures that link the shell of one cell to that of the next in the chain. This is a rare species, *Strangulonema barbadense*, known only from Upper Eocene deposits (40 million years old). This one is from an Indian Ocean core.

Many mineral specimens occur as large crystals of characteristic shape and colour. But many others that look granular or amorphous to the naked eye are in fact composed of microscopic crystals which may be just as regular in their form and are just as diagnostic of their species as the spectacular specimens prized by collectors. Electron microscopes can be used to study growth structures in minerals and, with the aid of X-ray detectors, their chemical compositions.

# Pyrite

*Crystals in a fossil ammonite* [x 1420]

Pyrite or iron pyrites, a form of iron disulphide, is the most abundant metallic sulphide mineral. It is a common component of fossils, replacing the original substance of shells etc., and is well known as the 'fool's gold' found in coal. In humid surroundings it may oxidize, producing sulphuric acid which can cause great damage to museum collections of fossils. The compact, intergrown, euhedral crystals seen in this specimen are usually stable.

*Infilling of a Cretaceous ammonite* [x 3030]

In this specimen the octahedral crystals are interspersed with roughly spherical aggregates of loosely-packed microcrystalline pyrite called framboids which are very unstable. In damp conditions the oxidation of the pyrites and production of acid may be accelerated.

# Calcite

*Scale from a kettle* [x 6710]

All water that has been in contact with the ground contains salts in solution. Spectrographic analysis of this sample, from a kettle used for boiling London tap water, showed the presence of calcium, iron, silicon, manganese, aluminium and magnesium. The infra-red spectrum showed the presence of calcite, a form of calcium carbonate, with trace impurities of silica and silicates.

*The surface of a rusty screw* [x 1360, insert x 730]

Rust, a mixture of hydrated oxides of iron, is a corrosion product forming on the surfaces of iron objects in the presence of moisture. The process is electrolytic in character and is accelerated by impurities in the iron and by dissolved gases such as carbon dioxide. The circular 'craters' appear to be at spots where water droplets have acted on the metal, inducing crystallization of the oxide to follow the contours of the droplet.